Teaching with Manipulatives:

Middle School Investigations

Rosamond Welchman

Vernon Hills, IL

Teaching with Manipulatives: Middle School Investigations
ISBN 0-938587-75-7
ETA 015067

ETA/Cuisenaire • Vernon Hills, IL 60061-1862
800-445-5985 • www.etacuisenaire.com

Printed in the United States of America.

07 08 09 10 11 12   10 9 8 7 6 5 4 3

Table of Contents

Introduction

How do you envision what students will take with them from your mathematics class—whether in memories, skills, or attitudes? We are all concerned that students master appropriate mathematical skills, but we also hope that they remember the excitement of mathematical investigations and that they are left with the disposition to pursue mathematics. The purpose of this book and its accompanying kit is to provide students in grades 5-8 with enjoyable and thought-provoking investigations that involve a variety of mathematics manipulatives and that support development of major curriculum topics.

When today's students become adults, they will be faced with complex problems requiring new and creative solutions. These future challenges have motivated new visions of classrooms where students develop mathematical power—the ability to use mathematics to formulate, investigate, and solve problems. The investigations presented here support these visions and help students develop confidence in their mathematical abilities. The activities give students many opportunities to tackle mathematical tasks, to reason about their ideas, and to communicate their thinking, both to teachers and to peers.

These new visions of student learning go far beyond what was measured in traditional standardized tests, which were based largely on memorization of computational procedures, vocabulary, formulae, and so on. Although those lower level thinking skills are perceived as important, a much richer mathematical experience involving higher order thinking skills such as reasoning, analysis, synthesis, and evaluation is seen to be essential. The National Council of Teachers of Mathematics has summarized and illustrated these trends in two documents: *Curriculum and Evaluation Standards for School Mathematics* (1989) and *Professional Standards for Teaching Mathematics* (1991).

The Manipulatives Needed

Manipulatives are an essential component of the investigations in this book. Owing to their inherent mathematical structure and visual appeal, they provide a natural context for posing interesting problems. When using manipulatives, students develop the confidence to experiment—an essential characteristic of good problem-solvers—because moving materials around is risk-free, leaving no trace, as does a pencil. Manipulatives allow students to represent their mathematical thinking through physical actions even before they can verbalize their reasoning. Students then use this physical expression of thought to develop abstract language and symbolism. Later, students can use manipulatives to play a different role, to verify the results of their more abstract thinking.

Introduction

The investigations require 12 different manipulatives—all of which can be found in the accompanying kit. The basic manipulatives are:

- *Geoboards*

  4, transparent and with an array of 5 by 5 pegs

- *Tangrams*

  4 sets, each of a different color

- *Polyhedra Dice*

  1 set of 6 pieces (tetrahedron, hexahedron, octahedron, decahedron, dodecahedron, icosahedron)

- *Standard Dice*

  6, 2 each of red, green, and white

- *Connecting Centimeter Cubes*

  200, 20 each of 10 different colors

- *Cuisenaire® Rods*

  2 sets, plastic

- *Two-Color Counters*

  100

- *Hundred Boards (0–99)*

  4, transparent

- *AlphaShapes™*

  52 pieces, 2 sets of 26 transparent shapes, 1 each in green and orange

- *Reflect-It™ Hinged Mirrors*

  4

The manipulatives used as support materials are:

- *Centimeter Grids*

  4, transparent

- *Color Chips*

  250, 50 each of 5 colors, transparent

- *Cuisenaire® Angle Rulers*

  4

- *Spinners*

  2, transparent

Each of these manipulatives is suitable for use by a small group of students. In addition, the transparent manipulatives are also suitable for use on an overhead projector with a whole class.

Contents of the Book

This book is divided into two sections. The 33 student activities, found at the back of the book, are the basis for the teacher notes that precede them. These student activities have been grouped according to manipulative, and are related to a range of mathematics topics, including measurement, geometry, algebra, probability, and number. Computations involving fractions, decimals, and percents are approached in the context of these mathematics topics.

Each activity sheet is meant to start students on an investigation of a broad idea by providing both a context and some tasks that launch students into the investigation. There are three to six activities for each manipulative. (Some activities will require more than one manipulative.) Most activities are in a 5-inch by 8-inch format and are presented two to a page so that they either can be cut apart and reproduced to form a set of task cards or can be duplicated on a transparency. Also included are blackline masters for two kinds of geoboard paper, isometric dot paper, and one-centimeter grid paper.

Each activity will most likely take at least a class period. More time will be required if all suggestions for follow-up activities are taken.

The teacher notes for each manipulative (or for related manipulatives, such as the Hundred Board and the Color Chips or the AlphaShapes, Angle Rulers, and Centimeter Grids) begin with an introduction to the nature and uses of the manipulative. Then, for each activity, there is a brief description of what students do and the mathematical potential of the investigation. This is followed by a reduced version of the student activity. Next, under the heading *In Preparation* are: what students need to know, directions for the teacher's assembly of supplementary materials, if needed, and occasional suggestions for introducing the activity. The rest of the notes are headed *Afterwards*. Here, in *Thinking Out Loud*, are suggestions for stimulating class discussion of the investigations, specifically, sample questions to clarify and extend students' thinking. Then, in *About Solutions,* is a discussion of mathematical ideas that are likely to arise during the investigation. Finally, under *Going Further*, are suggestions for extending each investigation, sometimes by having students design similar investigations with the same manipulative, sometimes by suggesting a similar activity with a different manipulative, and sometimes by encouraging students to generalize to a formula or an abstract representation.

Relating Investigations to the Curriculum

The grid on page 6 shows the mathematics topics that are the major emphasis of each investigation (denoted with "●") as well as the topics that arise incidentally (denoted with "○").

The investigations enable all students to have the intuitive experiences necessary for more successful formal development later on. By connecting

Introduction

mathematical ideas as they work, those students who already have been exposed to a particular topic in a structured way can find the investigations to be an "aha!" experience, whereas less experienced students can use the same investigations as an introduction to the topic. (In truth, many students in the middle grades have been exposed to more mathematical terminology than they have really understood or absorbed.)

It must be emphasized that these activities are not meant to be the only exposure that students have to the topics involved. Rather, they are intended to facilitate students' exploration with manipulatives in ways that usually may not be available to them.

Implementing the Investigations

If the manipulatives are new to students, allow time for them to explore each material before introducing the activity. Ask students what they notice about the material and how they think it could be used to develop mathematical ideas. Use the overhead projector to demonstrate or explain. Involve students by encouraging them to use it to share their thinking.

Choose those activities that support the mathematics you want your students to engage in. Then fit those activities into your classroom routines in any way that is comfortable for you. Students can work in small groups, as a whole class, or independently. Sometimes you may find one arrangement more suitable than another. For example, you might select a cluster of activities—perhaps three or four—relating to the same topic, and have students rotate—either in groups or individually—among these activities over several weeks. (Use the grid on page 6 to identify such clusters of investigations.) You might assemble the print materials and the related manipulatives into large, heavy-duty, zip-lock bags. These can be labeled and thumbtacked to a bulletin board. You could assign students to groups, having as many groups as activities. If you prepare a chart with the group names and activities, students can manage the distribution and record-keeping for the activities themselves. Or you could assemble enough additional material so that the entire class can work on the same investigation in small groups.

As students work, they should respond to the questions on the activity cards either in a notebook or on a duplicated record sheet. This written work supports students' thinking and helps them to extend their reasoning and to make whole-class presentations. It can also be used by you for assessment purposes. However, since oral communication is a vital prerequisite for written communication, all students should have the opportunity to talk about their ideas before they try to write them down.

Regardless of how you have implemented an investigation, once all students are finished working, a whole-class discussion is invaluable, since it gives students an opportunity to reflect on what they have done and to deepen their mathematical understanding.

Ideas for Assessment

As students work, circulate and listen. Make note of the vocabulary students use, the approaches they try, and the observations they make to one another. Question groups about their thinking when the opportunity arises.

On the activity sheets, students are encouraged (either explicitly or implicitly) to do some form of written record-keeping. This can be extended to writing in a math journal. The activities and materials should give students many ideas for entries. Results of the investigations could also provide suitable entries in an individual or class mathematics portfolio—a collection of exemplary mathematics work.

Topic Grid

	Ratio, proportion	Fractions	Decimals, percents	Number theory, factors and multiples	Computation and estimation	Patterns and functions	Algebra	Sampling, graphing	Probability	Spatial visualization	Properties of polygons	Properties of circles	Similarity, congruence	Symmetry	Pythagorean Theorem	Solid geometry	Angle measurement, relationships	Length	Area, perimeter	Surface area, volume, mass
Mystery Shapes										●			○			○	○	○		
Patterns with Area				○	●	○													●	
Squares on a Triangle					○	○				○			●					○	●	
Tangrams and the Geoboard	○	●			○					○								○		
Similarity and Triangles	○	○			○					○	○		●			○	○	○		
Shopping for Tangram Pieces	●	○	○	○	○	○				●								○		
Describing Dice					●	○										●				
Rolling Threes (game)	○		○	○	○			○	●							○				
Two Dice or One?	○		○					○	●							○				
Making the Most of It (game)	○			○	●			○	○											
Any Order (game)					●	○	●		○											
Building with Cubes										●				○						○
Patterns with Cubes	○			○	●	○				●		○								●
The Price of Fruit	●	○	○		●			○												●
Perimeters				○	○	○				○	○								●	
Rod Sculptures					○	○				○						○				●
Fraction Rod Race (game)		●			○				●								○			
Tossing Four Counters	○							●	●											
Arranging Four Counters	○				○			●	●											
Fraction Flips	○	●			○		○													
Collecting Counters (game)					○	○	●		●											
Chip Patterns on a Hundred Board				○	●	●	●													
Chips in a Row (game)		○			●				○											
Cover by Clues (game)				●	○	●	●													
Labels on Loops (game)										●	○	○	○				○	○		
Properties of Shapes										●		○	○				○	○		
Ways to Find Area		○			○					○									●	
Tracing Angles					○	○	○			○							●			
Tessellations										●							○	○		
Circles	○		○		○	○	○					●						●	●	
Ways to Measure Angles					○	○				○							●			
Lines of Symmetry				○	○					○				●						
Finding Shapes									○	●				●						

● means major emphasis
○ means less emphasis

Geoboards

 Geoboards are square boards on which pegs are arranged at regular intervals. Students can easily stretch rubber bands from peg to peg to create polygons. They can also change these shapes quickly and effortlessly into a wide variety of other shapes. As a result, students find it exciting to use geoboards to experiment with geometric ideas.

The transparent geoboards in the kit have a 5 by 5 array of pegs. The unit for length on a geoboard is the shortest distance betweeen two pegs, and the unit for area is the smallest square whose corners are pegs. Geoboards can be used to pose many interesting challenges related to measurement of length and area.

Using the transparent geoboards, students can share their work on an overhead projector. The transparent nature of these geoboards also allows students to verify congruence of lengths, angles, and shapes by putting one geoboard on top of another.

Three geoboard activities and teacher notes follow. (See also Tangrams and the Geoboard, *in* **Tangrams**, *in which a geoboard is used together with a set of tangram pieces.)*

Mystery Shapes

Students create geoboard shapes that match sets of clues. This allows students to become familiar with the geoboard and provides them with experience in recognizing and applying geometric properties.

As students change the positions and shapes of the rubber bands to match the clues, they increase their facility with using geometric language and build mental images of polygons. Because a shape that satisfies one clue may not satisfy the next clue, students are often required to adjust their shape. As they reevaluate their thinking, they are developing logical reasoning skills.

Mystery Shapes

You will need a geoboard, rubber bands, and sets of clue cards.

1. What is the mystery shape? Read all the clues, then use your geoboard to identify the shape.

> *Clue Set 1*
> It has four sides.
> All of its angles are congruent.
> One side is twice as long as another side.
> The rubber band does not touch the center peg.

What shape did you find? Can you find another shape to fit these clues? Record your solutions.

2. Find another mystery shape. Use another set of clue cards. Decide whether to work alone or with three classmates. If you work in a group, each person should

- have a geoboard and one clue.
- read his or her clue aloud.
- make a shape that satisfies each clue.
- touch only his or her own geoboard.

3. Repeat this activity. Use another set of clue cards.

4. Make a new mystery shape. Write four clues for it. Have others use the clues to make four different shapes. Add your clues to a class set.

Geoboards

In Preparation

Students may or may not be familiar with the vocabulary used in the clues. Such experience is helpful but not essential for this activity. As students work, urge them to consult a dictionary or mathematical glossary for any terms they don't understand. If students work cooperatively, the group may well be able to fill in gaps in a particular student's knowledge.

You will need to prepare sets of clues such as the ones below, which you will find ready for reproduction on page 89.

Clue Set 2	Clue Set 3	Clue Set 4
It has a right triangle.	It has four sides.	It is not convex.
It has one interior peg.	It has no line of symmetry.	It has five sides.
It is isosceles.	Its area is six square units.	It has no interior pegs.
It uses a corner peg.	Just one pair of opposite sides is parallel.	It has a right angle.

If students choose to work in a group of four, cut apart the clues and keep each set in its own envelope.

Afterwards

Thinking Out Loud

As part of the discussion, use prompts such as these:

- Describe your thinking as you worked with each set of clues.
- What different solutions did you discover for each set of clues? How are these solutions alike? different?
- Would you have more solutions if you omitted a clue?
- Could you write more clues that would apply to all your solutions?

About Solutions

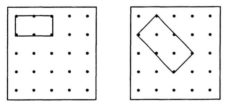

In Clue Set 1, the first and second clues guarantee that the shape is a rectangle. There are three possible sizes for a rectangle with sides in the ratio of 2 to 1 that will fit on a 5 by 5 geoboard. All of these can be arranged to avoid the center peg.

In Clue Set 2, the solution must be a right isosceles triangle of one or two sizes. Putting the two transparent geoboards on top of each other shows that the two triangles aren't congruent. (You might also use the Pythagorean Theorem to find the lengths of the sides of each triangle.) The shorter side of one is 3 units, whereas the shorter side of the other is $\sqrt{8}$ units, or approximately 2.8 units.

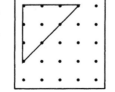 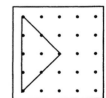

In Clue Set 3, the first and fourth clues could be replaced by the single clue "It is a trapezoid." Here, there are four different solutions, each of which can be in different positions on the geoboard.

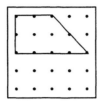 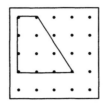 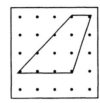 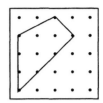

Clue Set 4 also has several different solutions.

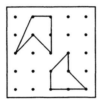 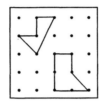 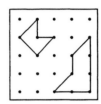

Students may realize that if they find one solution, they may be able to transform it in some way to get a different one.

Going Further

Students might write clue sets that use other manipulatives, such as tangrams or Cuisenaire Rods, to build the mystery shapes.

Patterns with Area

Students find the area of shapes that they construct from specified numbers of interior and boundary pegs.

In addition to using an important problem-solving technique—keeping one variable constant while changing another—students learn about Pick's Theorem. Pick's Theorem describes the relationship among the numbers of boundary pegs (B), interior pegs (I), and area in geoboard units (A) for any polygon made on a geoboard. Students may be surprised to discover that if any two shapes have the same number of boundary pegs and interior pegs, the areas of these shapes will be the same. Students may well then discover how they can find (A)—area—if they know the values for (B) and (I). Using a geoboard facilitates the discovery of these relationships, since students can easily manipulate the rubber bands to create a wide variety of shapes with specified characteristics.

This activity also provides students with an opportunity to understand how algebraic symbolism helps clarify mathematical ideas and simplifies communication.

Geoboards

Patterns with Area

You will need a geoboard and geoboard dot paper.

The pentagon on the geoboard has one interior peg, six boundary pegs, and an area of three square units.

Make other shapes that have one interior peg and six boundary pegs. Find the area of each shape.

Record your results on dot paper. Write about what you notice.

Make four more shapes that have one interior peg but different numbers of boundary pegs. For each shape, find the number of boundary pegs, (B), and the area, (A). Record your results.

How do your results compare with those of your classmates?

What patterns can you find?

Here are some ideas for finding the area of your geoboard shapes.

| a | b | c | d |

a. Count squares or half squares. (4 cm²)

b. Find the area of a right triangle by surrounding it with a rectangle that has twice its area. Then take half of the area of that rectangle. (3 cm²)

c. Combine the first two methods. (5 cm²)

d. Surround the triangle with a rectangle. Find the area of the rectangle and each right triangle. Subtract total triangle area from the rectangle. (3.5 cm²)

In Preparation

Students should have had some experience with the techniques for finding area that are included on the student activity page.

Afterwards

Thinking Out Loud

As part of the discussion, use prompts such as these:

- Explain how you worked.
- How could you change a shape without changing the number of interior pegs?
- What patterns did you notice?
- How did you record what you found?

About Solutions

Pick's Theorem states that the area of a polygon can be found by taking half the number of boundary pegs, adding the number of interior pegs, and subtracting 1. In symbols, if (B) is the number of boundary pegs, (I) is the number of interior pegs, and (A) is the area in geoboard units, then for any polygon made on a geoboard

$$A = 1/2\ B + I - 1.$$

Often, students find it easier to discover this relationship if it is broken down into special cases, as happens in the student activity, when students first consider a situation in which there is just one interior peg. For example,

$$\text{when } I = 0, A = 1/2B + 0 - 1 = 1/2B - 1$$
$$\text{when } I = 1, A = 1/2B + 1 - 1 = 1/2B$$
$$\text{when } I = 2, A = 1/2B + 2 - 1 = 1/2B + 1$$

Students may describe these patterns in words or in algebraic language.

Going
Further

Once students think they have discovered Pick's Theorem, you might show them an example such as this, which is not a polygon.

Help students to see that the formula does not apply in this case. Have students try to modify the formula to suit shapes like these, with "holes."

Squares on a Triangle

Students use the geoboard and centimeter dot paper to construct triangles and then build squares on each side. Students go on to find the areas of these squares and to search for relationships.

By making triangles, building squares on each side, and looking for relationships among the areas of the squares, students are led to formulate the Pythagorean Theorem in the way that it was seen by the ancient Greeks—as a relationship among the areas of squares built on the sides of right triangles. This theorem is named after

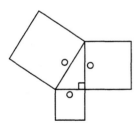

Pythagoras, a mathematician who is thought to have lived from 569 to 500 B.C. (although the theorem was known to Babylonian mathematicians more than a thousand years before Pythagoras was born). Investigating right triangles as well as nonright triangles makes it clear that the Pythagorean Theorem applies only to right triangles. By using concrete materials, students are able to truly focus on the concept rather than on the manipulation of symbols or on a formula.

Squares on a Triangle

You will need a geoboard and geoboard dot paper.

1. Make as many squares of different sizes as you can on the geoboard. Record them on dot paper.

2. Find the area of each square.

3. Make this right triangle on your geoboard or on dot paper. Build squares on each side of the triangle and find their areas. Record the sum of the areas of the two smaller squares.

4. Repeat this procedure for the two triangles below.

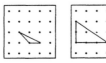

What do you observe about the sum of the squares built on the two shorter sides of a right triangle and the area of the square built on the longer side? Write down an idea that might always apply.

Geoboards

In Preparation

Students should be somewhat familiar with the techniques for finding area that appear on the *Patterns with Area* student activity page.

Afterwards

Thinking Out Loud

As part of the discussion, use prompts such as these:

- How do you know that you have made squares?
- What methods did you use to find the area of the squares?
- How did you record your findings?
- What patterns did you discover?

About Solutions

There are eight different sizes of squares that can be made on a 5 by 5 geoboard with areas, respectively, of 1, 2, 4, 5, 8, 9, 10, and 16 square units. If a, b, and c are the lengths of the sides of a right triangle, and c is the length of the side opposite the right angle (the hypoteneuse), then the areas of the three squares are a^2, b^2, and c^2, and students will discover that $a^2 + b^2 = c^2$. Students will find that this relationship does not hold for nonright triangles.

Going Further

Students can be guided to discover that they can find the length of any line segment on the geoboard or on centimeter grid paper by constructing a square with the segment as its side, finding its area, and taking the square root of that area. Students can discover that the lengths of the sides are the square roots of these numbers, namely $1, \sqrt{2}, 2, \sqrt{5}, 2\sqrt{2}, 3, \sqrt{10}$ and 4.

Tangrams

The tangram is a deceptively simple puzzle with rich mathematical potential. Each tangram set consists of five triangles, one square, and one parallelogram. The pieces in each tangram set in the kit all fit together to form a 10-centimeter square as well as an amazing variety of other shapes. There are four tangram sets in the kit, each a different color.

Tangrams are sometimes used for puzzles in which students arrange the pieces to fill a specific outline or to duplicate an arrangement from a picture. This kind of puzzle helps students to recognize similarities and to perform spatial tasks with greater facility.

Tangram pieces are designed to have certain geometric relationships: Each piece can be constructed from one, two, or four copies of the smallest triangle. For this reason, by using tangram pieces to measure area, students are able to develop an understanding of fractions and ratios.

Three tangram activities and teacher notes follow.

Tangrams and the Geoboard

Students use a geoboard to construct a shape that is designated as the unit and then discover what fraction of this unit is represented by each tangram piece. They see that when the shape designated as the unit is changed, each particular piece will stand for a different fraction.

The flexibility of the rubber band makes it easy for students to explore the effects of changing the unit when assigning fractional values to the tangram pieces. Furthermore, it is possible to construct congruent copies of all tangram pieces because the geoboard unit for length (the shortest distance from peg to peg) is the same as the length of the short side of the smallest tangram triangle.

Tangrams and the Geoboard

You will need a transparent geoboard, a rubber band, and one set of tangram pieces.

1. Make a square like the one to the right. Turn the geoboard over. Answer the following question by placing tangram pieces on top of the geoboard:

   What fractional part of the square is each of the tangram pieces?

   Trace each piece. Record the fraction inside and write a convincing argument explaining why your answer is correct.

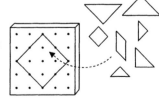

2. Make a rectangle like the one to the left. What fractional part of the rectangle is each of the five different pieces?

3. The small triangle has a different value when used in the square than it does when used in the rectangle. Why?

4. Make a shape of which the large triangle is 1/2. What fractional part of this shape is each of the other four pieces? Write about your strategies for answering this question and any patterns that you notice.

Tangrams

In Preparation

Since this activity shows how area relationships among tangram pieces can illustrate concepts of fractions, students should have had prior experience with tangrams, for example, as described in the *In Preparation* section under *Shopping with Tangram Pieces*.

Afterwards

Thinking Out Loud

As part of the discussion, use prompts such as these:

- If you know how many large triangles fit in the square, can you tell, without actually counting, how many small triangles would fit? How?
- How can you figure out how many small tangram squares would fit in the rubber band square?
- How can you make other shapes of which the large triangle is one quarter? Would each tangram piece be the same fractional part of these shapes as it is of the square?
- How do the tangram pieces show you what 1/4 of 1/2 is?

About Solutions

There are enough large triangles in four sets of tangrams to fill up the square on the geoboard so that it can easily be seen that the large triangle represents 1/4. Because there are not enough of the medium triangles to fill up the square, however, the following must be reasoned: Since the medium triangle is 1/2 of the large one, 8 will fill the square. This means a medium triangle represents 1/8. What this says mathematically is that 1/2 of 1/4 is equal to 1/8, or written symbolically, 1/2 × 1/4 = 1/8. Similarly, the small triangle represents 1/16.

The unit is changed in part 2 so that the large triangle represents 1/3, which now makes the medium triangle represent 1/6—that is, 1/2 of 1/3 is equal to 1/6.

Going Further

Students can be challenged to fill in the geoboard outlines with pieces of different colors and identify what fractional part each outline is of each color. Results can be added to a class puzzle file. For example, you might ask students to fill in the square in part 1 with 7 pieces so that 1/2 of the area is red, 1/4 is blue, 1/8 is green (a). To solve this, students may reason that since 16 small triangles fill the square, to make 1/2 of the area red, the equivalent of 8 small triangles must be red, and so on.

Such questions can also be posed using percents, or shapes, instead of colors. You might ask, for example, whether students can fill in the square so that 25% of the area is made up of parallelograms, 50% of triangles, and the rest of squares (b).

a   b

To assess students' understanding of how choice of unit affects the fractional value assigned to a tangram piece, ask them to make a shape on the geoboard that can be used to show that $2/10 = 1/5$.

Similarity and Triangles

Students construct a number of triangles from the tangram pieces—all of which must be similar, owing to the nature of the pieces—and explore ways in which they are the same and different. Students also construct parallelograms that are not similar. Construction of similar shapes with the manipulatives build spatial visualization skills in a problem-solving context. Through collection of data about angle, length, and area of these similar and nonsimilar shapes, students deepen their understanding of similarity.

Similarity and Triangles

You will need a set of tangram pieces, an angle ruler, a transparent centimeter grid, and a calculator.

1. Take one of each of the three different sizes of triangles. Make a fourth triangle with the remaining pieces: large triangle, small triangle, square, and parallelogram. Arrange the four triangles in front of you.

2. Compare the triangles in as many ways as you can. (You can measure them with the angle ruler or the transparent centimeter grid.) Find how the triangles are the same and how they are different. Record your findings.

   Mathematicians would say that the four triangles in front of you are similar to each other, but none of them is similar to the one on the right. Why do you think this is so?

3. Take the parallelogram, then use the other tangram pieces to make two more parallelograms, one using the two large triangles and the other using the remaining pieces (the square, the two small triangles, and the medium triangle). Arrange the three parallelograms in front of you, as shown.

   Mathematicians would say that the top and bottom parallelograms are similar to each other but neither is similar to the one in the middle. Why do you think this is so?

4. What *are* similar shapes? Write a definition to help someone understand.

In Preparation

This activity requires measuring length in centimeters (to the nearest tenth of a centimeter), area in square centimeters (by counting squares in a grid and using relationships among tangram pieces), and finding ratios among these measures. Therefore, students should have had previous experience with measuring in these metric units.

Tangrams

Afterwards

Thinking Out Loud

As part of the discussion, use prompts such as these:
- If these shapes were photographs, could one be an enlargement of another?
- In what ways did you compare the two triangles?
- Is there a triangle that is twice as big as the small one? What is meant by "twice as big?"
- Why weren't all parallelograms similar to one another?

The ratio between corresponding sides of similar figures is always the same. The similar triangles made from tangram pieces have the following relationships: When one side is doubled in length, the other sides are also; and when one side is doubled in length, the area is multiplied by 4. More generally, when a side is multiplied by r, all other sides are also multiplied by r, but the area is multiplied by r^2.

Since the only angle measures represented in the set of tangram pieces are $45°$, $90°$, and $135°$, the only triangles that can be formed from tangram pieces are right isosceles triangles, with two $45°$ angles. The ratio between short and long sides of these similar triangles is always the same (actually $\sqrt{2}$, or about 1.4). Working with the parallelograms deepens understanding of what it means for figures to be similar; it is not enough only to have congruent angles. The drawings of a rhombus and a square to the left, in which the sides of one are double the sides of the other, reveal that having sides in the same ratio is not enough. For two shapes to be similar, corresponding angles must be congruent and also corresponding sides must have the same ratio.

Going Further

In this activity, students made triangles of four different sizes. Now ask students to find triangles of other sizes using just one set of tangram pieces. They might explore this by doubling, which would result in five triangle sizes, plus a sixth that is not an exact double.

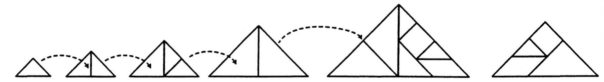

Encourage students to explore other shapes by making pairs of similar figures and verifying any patterns they notice. Students might start with a simple two-piece tangram construction and use tangram pieces to enlarge it. In the process they may discover, for one thing, that all squares are similar but not all rectangles are.

Students can extend this investigation to other materials. If you have access to a photocopier that enlarges or shrinks, make several copies of a given shape, all of different sizes, and ask students to measure angles, lengths, and areas to verify their results. At another time, give students several shapes that are similar and one that is a "near miss"—a shape that looks like it might be similar to others. Challenge students to find the shape that was not made from your original by the photocopier.

Shopping for Tangram Pieces

After seeing the "prices" of a few sample tangram pieces (which are determined by color and area), students assign prices to all the other tangram pieces in the four sets. Students then fill in puzzles with pieces of a particular price, thereby combining a spatial challenge with a numerical one. In addition, in assigning prices, students experience ratio, since a piece twice as big as another will be valued at twice the price.

Manipulating the pieces allows students to easily explore different combinations and to express their thinking physically.

Shopping for Tangram Pieces

You will need a set of tangram pieces.

Suppose the manufacturer of the four tangram sets must buy each color of dye at a different price. So, both color and size will determine the cost of each piece. For example, because the square can be made from two small triangles, the red square will cost twice as much as the small red triangle.

1. Use tangram pieces to figure out the cost of the pieces listed on the chart. Copy the chart and record your answers. List any patterns you notice.

	small triangle	medium triangle	large triangle	square	parallelogram
red	5¢			10¢	
green		12¢			
blue			16¢		
yellow				6¢	

2. Fill in the outline with red and yellow pieces so that the shape will cost 56 cents. In writing, explain the process you went through to solve this problem.

3. Make your own tangram shape. Trace its outline. Figure out what your shape costs. Then tell someone the amount and have that person find the colors and pieces that will complete your outline and provide prices for each piece.

Tangrams

In Preparation

Because this activity builds on the spatial relationships among the pieces, students will benefit from previous experience with tangram puzzles. If you wish, display a couple of outlines for students to fill in exactly with the tangram pieces. Then you might have students make outlines of their own for their classmates to try.

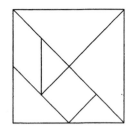

Students can learn a good deal about the structure of the tangram pieces by making their own sets from construction paper before using the tangrams in the kit. (Instructions for making tangrams are given in several resource books in Cuisenaire's catalog.)

Afterwards

Thinking Out Loud

As part of the discussion, use prompts such as these:

- How is the small triangle related to the medium one? to the large one?
- How much would it cost to fill the outline with red pieces? with yellow pieces? What would happen if you changed the color of just one piece?
- What is the fewest number of pieces that you can use to fill in the outline? What is the greatest number of pieces? What is the least it could cost if you used different colors? What is the most?
- If you know the total cost of all five pieces, how would you find the cost of each individual shape?

About Solutions

The medium triangle, the square, and the parallelogram all can be made from two small triangles; therefore, they all will have the same price—twice that of the small triangle. The large triangle can be made from four small triangles and thus will cost four times what the small triangle costs. These ratios are essential to figuring out the cost of each tangram piece.

The puzzle in part 2 can be filled in by using all the pieces except the medium triangle. This can be explained by comparing the diagram to a square made with all seven pieces. Hence, if the puzzle is made in red, it will cost $5 + 5 + 20 + 20 + 10 + 10 = 70$ cents. To find the costs of the puzzles made in other colors, students could do a similar addition, or they might apply a ratio: The small green triangle costs 6 cents, so the entire puzzle in green would cost $6/5 \times 70 = 84$ cents.

Here is one general approach to solving part 2. Imagine the puzzle filled in entirely with small triangles, which are either red, costing 5 cents, or yellow, costing 3 cents. The puzzle requires 14 small triangles to be used and the totals are 70 cents and 42 cents, respectively, if only one color is used. Experiment by replacing the triangles with pieces of the other color, one at a time, until arriving at a total of 56 cents. You might make a chart and find a pattern—each time a small yellow triangle is replaced by a red one, the price increases by 2 cents. If seven pieces are red (costing 35 cents) and seven are yellow (costing 21 cents), the total cost will be 56 cents.

Still another way to tackle this problem is by trial and error. Start with a number of reds, say five (which cost 25 cents). This would mean that there will be $14 - 5 = 9$ yellows (costing 27 cents), which would make a total cost of 52 cents. Since this price is too low, increase the number of red pieces and decrease the number of yellow pieces.

Dice

The kit includes six dice with 4, 6, 8, 10, 12, and 20 faces, all labeled with numerals, and six standard cube dice in three colors (with 1 to 6 dots on each face). The dice include all of the five possible regular polyhedra—that is, those polyhedra whose faces are regular polygons (all faces congruent and all angles congruent) and that have the same number of faces meeting at each vertex. These are called a tetrahedron (4 faces), a cube or hexahedron (6 faces), an octahedron (8 faces), a dodecahedron (12 faces), and an icosahedron (20 faces). The remaining die, a decahedron (10 faces), is not regular because its faces are kite-shaped polygons.

All the dice are fair dice. This means that when rolled, every number or face has an equal chance of being selected.

The interesting variety of solids represented by the polyhedra dice invite exploration of their geometric aspects, such as the shapes of faces and numbers of faces, edges, and vertices. This variety also allows many experiments with the dice that help students understand concepts of probability, such as *sample space* and *equally likely events*.

When students try the games that follow, they should note that the die that *theoretically* should win does not always win but, over the long run, it is more likely to. It is important for students to have many experiences in probability in which they compare theoretical results with the results of repeated experiments. These will help students to learn that the more trials there are, the more it is likely that experiments will match theory.

Finally, the fact that all of the polyhedra dice have numbers on their faces makes them useful for choosing numbers in problem-solving activities involving computation.

Five dice activities and teacher notes follow.

Describing Dice

Students look at the polyhedra dice as geometric solids and find ways in which they are the same and different. In the process, students look for patterns as they count numbers of faces (the flat surfaces), edges (the line segments that border the faces), and vertices (the points where the edges meet). This activity allows students to become familiar with the dice and also encourages the use of appropriate geometric vocabulary. Students need to see and handle three-dimensional models to give life to the two-dimensional representations of polyhedra that they see on a printed page.

Students can also generalize from the patterns they find to formulate a well-known mathematical relationship, Euler's Formula.

Describing Dice

You will need a partner and a set of polyhedra dice.

1. Pick two polyhedra dice. Make a list of ways in which the two dice are alike. Make another list of ways in which they are different. Try this again for another pair of polyhedra dice.

2. One way to describe dice is to count aspects of them. You might count the faces (the flat surfaces), the edges (the line segments where two faces meet), or the vertices (the points where edges meet). Make a table showing the numbers of faces, vertices, and edges of each die and describe any patterns that you find.

3. For each die, write a description that applies only to it and not to any other die. Can you do this in more than one way? For example, "the die that has 3 triangular faces meeting at a vertex" describes the tetrahedron die. Two other unique descriptions would be "the die with 3 numbers to each face" and "the die with 6 edges."

In Preparation

Students will benefit from some familiarity with the properties of various solids. They might sort those solids with flat faces (polyhedra) from the others—for example, spheres, cones, and cylinders. Introduce students to the meaning of *face*, *edge*, and *vertex*. If necessary, use larger models—for instance, cardboard boxes or wooden blocks.

Afterwards

Thinking Out Loud

As part of the discussion, use prompts such as these:

- What are the shapes of the faces of the dice?
- If you made a structure like this, with toothpicks for the edges and joined by miniature marshmallows at the vertices, how many toothpicks and marshmallows would you need?
- What is the difference between a square and a cube?
- What happens when you add the numbers of faces and vertices and compare the sum to the number of edges?
- Would any of your patterns work for another solid such as a triangular prism?

About Solutions

Among the many ways to compare two dice are color, size, shape, number of faces, edges, and vertices, and the numerical labeling of each face. The following table indicates the number of faces, edges, and vertices for each of the six die.

	Faces (F)	Edges (E)	Vertices (V)
Tetrahedron	4	6	4
Cube	6	12	8
Octahedron	8	12	6
Decahedron	10	20	12
Dodecahedron	12	30	20
Icosahedron	20	30	12

Dice

This table contains a number of patterns, such as the following:
- the number of faces, vertices, and edges are all even.
- for the pairs of dice that have the same number of edges, the numbers of faces and vertices are interchangeable.
- the sum of the numbers of faces and the vertices is always two more than the number of edges.

The third pattern above can be expressed algebraically as $F - E + V = 2$, and is known as Euler's Formula. This relationship is true for any polyhedron, not just those listed on the table.

There are many possible unique descriptions of each polyhedron die. Here are a few examples:

Cube	The die that has some pairs of perpendicular edges.
	The die that has 4 parallel edges.
Octahedron	The die with 6 pairs of parallel edges.
	The die with twice as many edges as vertices.
Decahedron	The die on which the sum of the numbers on opposite faces is always 9.
	The die whose faces are not regular polygons.
Dodecahedron	The die with the same number of faces as months of the year.
	The die with pentagonal faces.

The pairing of the cube with the octahedron, the dodecahedron with the icosahedron, and the tetrahedron with itself corresponds to the fact that a similar copy of each polyhedron can be fitted inside its pair, with a vertex of one in the middle of the face of another, as shown below.

Suggest that students make their own models of these solids from light cardboard. Students can make the faces by tracing around AlphaShapes. They can then arrange the shapes on card stock with the faces next to each other so that edges can be obtained by folding.

Rolling Threes

Students predict which of the polyhedra dice will be the first to produce three 3's if all are rolled repeatedly. Students then roll all of the dice and compare their results with their predictions. Students develop an understanding of probability as they list all possible outcomes in the sample space and compare empirical data to theoretical data. They come to realize that since all faces are equally likely to be rolled, the more faces a die has, the less likely it is that any one number will be rolled.

In a second game, students predict which dice will be the first to produce three multiples of 3. This situation differs from the first because there are more chances for success here, since most dice have more than one multiple of 3. In addition, when comparing probabilities, students have the opportunity to see the advantage of using decimals or percents instead of fractions.

Rolling Threes

These are two games for two to six players. You will need a set of polyhedra dice. Share the dice so that each person is responsible for one to three of them.

Game 1   All of the dice are rolled at the same time, over and over. The winning die is the first to land on 3 three times.

- Predict which die you think will win.
- Play this game at least three times or until you feel sure about what the results will be any time you play. Keep a record of your results for each game.
- Discuss your results.
- Write about why your predictions did or did not match your results.

Game 2   This is the same as *Game 1* except that a die wins if it is the first to land on *multiples* of 3 three times.

- Predict which die you think will win.
- Write about how you made your prediction.
- Play this game as before.
- Compare your results to your prediction.

Dice

In Preparation

Students will find it helpful to have some experience with describing sample spaces and the notion of "fair" versus "unfair." If students repeatedly roll a standard die or toss a coin, they will readily understand that *fair* means that over the long run each number or outcome will be rolled *approximately*, but not *exactly*, the same number of times.

Afterwards

Thinking Out Loud

As part of the discussion, use prompts such as these:
- What kind of a record did you keep of your results?
- Did your results match your prediction?
- Which die, if any, has a better chance of producing a multiple of 3? Why?
- Do you think that if you did the activity again, your results would match your prediction? Why?
- How could you check to see if a die is "fair?"

About Solutions

There are different numbers of possible outcomes for each die when it is tossed. For any one die, all of these are equally likely. Expressed in more formal language, there are different "sample spaces." Since there is just one 3 on each die, the fewer faces there are, the more likely one is to roll a 3. One can say that the probability of rolling a 3 will be (in order of the number of faces of the dice) 1/4, 1/6, 1/8, 1/10, 1/12, and 1/20. Thus, in *Game 1*, the tetrahedron die, which has the least number of faces, should win. This is not true, however, for *Game 2*. Counting the number of multiples of 3 on each die shows that the probability of rolling a multiple of 3 is (again in order of the number of faces of the dice) 1/4, 2/6, 2/8, 4/10, 4/12, and 6/20 (note that 0 is a multiple of 3). If these fractions are converted to decimals or percents, they are easily compared, showing that the 10-faced dice should win in *Game 2*.

Going Further

To assess understanding of this activity, ask students to predict, and then experiment to find out which die would be the most likely *not* to roll a 2 for the greatest number of rounds. Here, one wants the probability of rolling a 2 to be as small as possible; the 20-faced dice is most likely to avoid a 2.

Students can be asked to design a die that selects one of the numbers 1-5. Ask whether they expect such a die to be fair. Then have students make a model and test it. (Students may realize that they could, in fact, use either the icosahedron or the decahedron, by renumbering faces.)

The 6-faced cube is commonly used to pick numbers in games. Ask students what would happen if they used 6-faced solids other than a cube, for example, small boxes, erasers, or blocks. Students might find or make some models, number the faces, and predict what percent of the time they will get each number if they roll them 20 times. They should then do so and keep a record of results. If students experiment with tossing noncubical boxes, they will gain valuable insight into the limitations of theoretical models in probability.

Two Dice or One?

Students compare their chances of getting a 3 when rolling just the 10-faced polyhedron die to their chances when rolling the cube and the 8-faced die together and taking their sum. As in *Rolling Threes*, by rolling the dice, students learn more about how theoretical probability relates to experience.

This time, students consider an even larger sample space consisting of all the possible ways in which two dice can be tossed. Students also consider the probability of an event that can occur in more than one way (that is, the 3 can be rolled in two different ways with the two dice.)

As in *Rolling Threes*, when comparing probabilities expressed as fractions, students can see the advantage of converting fractions to decimals or percents.

Two Dice or One?

You will need a partner and a set of polyhedra dice.

Suppose you will win $100 dollars if you roll a 3! You have a choice of which dice to roll:

- just the 10-faced die *or*
- both the 8-faced die and the cube die, where you take the sum of the numbers rolled.

Which would each of you choose? Explain why.

Now, you and your partner should each roll one of the two choices 20 times. Which of you got a 3 more often? Did this agree with your choice?

Compare your results with those of your classmates.

In Preparation

Students should have had some experience with equally likely events as well as with comparing actual results to predictions, as in *Rolling Threes*.

Afterwards

Thinking Out Loud

As part of the discussion, use prompts such as these:
- Are all numbers equally likely to be rolled with the 10-faced die? Why?
- How could you get a sum of 3 from the pair of dice? How could you get a sum of 9?

- What sums can be rolled with the pair of dice? Are all sums equally likely to be rolled? How do you know?
- How could you list all the possible ways you could toss the two dice?
- Suppose you could win 100 dollars if you rolled a 3, and you could pick any single die or any pair of dice. What would you choose? Why?

About Solutions

With the 10-faced die, you have one chance out of ten to roll a 3, so the probability of rolling a 3 is 1/10, or .1. It is harder to find the number of possible rolls of the pair of dice. For these, all possible rolls can be shown in a chart like the one below.

There are 6 × 8 = 48 ordered pairs, each of which represents a roll. As an example, (2,1) would represent rolling a 2 on the octahedron die and a 1 on the cube. The pairs whose sum is 3 are circled. There are only two of these; therefore, the probability of getting a sum of 3 is 2/48, or about .04, less than .1.

If the object of the game were to roll a 9, the probability of doing this would still be 1/10 with the 10-faced die. The pairs in the chart whose sum is 9 are marked with a square. Since there are six of these, the probability of rolling a sum of 9 is 6/48 = 1/8 = .125, or more than .1. This means that if rolling a 9 meant winning 100 dollars, someone would do better to choose the pair of dice than the single die.

		cube				
	1	**2**	**3**	**4**	**5**	**6**
1	(1,1)	((1,2))	(1,3)	(1,4)	(1,5)	(1,6)
2	((2,1))	(2,2)	(2,3)	(2,4)	(2,5)	(2,6)
3	(3,1)	(3,2)	(3,3)	(3,4)	(3,5)	[(3,6)]
4	(4,1)	(4 2)	(4,3)	(4,4)	[(4,5)]	(4,6)
5	(5,1)	(5 2)	(5,3)	[(5,4)]	(5,5)	(5,6)
6	(6,1)	(6,2)	[(6,3)]	(6,4)	(6,5)	(6,6)
7	(7,1)	[(7,2)]	(7,3)	(7,4)	(7,5)	(7,6)
8	[(8,1)]	(8,2)	(8,3)	(8,4)	(8,5)	(8,6)

(octahedron labels rows 1–8)

Going Further

Ask students to design similar activities using other combinations of dice or other target numbers. One challenge would be to have them describe a rule for winning whereby the two choices of dice yield exactly the same chances of winning.

As a modification of the game on the activity sheet, have students take the product of the numbers of the two dice (if they choose two dice), and win the money if they roll an even number. Before students begin, ask which they would choose to roll, the two dice or one, and have them explain why. Suggest that they roll each of the choices 20 times and tally their results. The only way to get a product that is not even is to get two odd numbers. Using the octahedron and the cube, there are 12 such pairs, as can be seen in the chart above; the other 36 pairs all give an even product. Therefore, the probability of getting an even product when the two dice are rolled is 36/48 = 3/4. For the 10-faced die, the probability of getting an even number is 5/10 = 1/2.

Making the Most of It

Students try to make the greatest possible product by filling in blanks on a recording sheet with numbers rolled on a die one at a time. Students can choose from among four polyhedra dice, and may come to use probability theory to make decisions.

This activity builds on students' understanding of the role of place value in computation and provides experience in doing mental computation.

Making the Most of It

This is a game for two to six players. You will need only the polyhedra dice with single-digit numbers. Use a calculator if you wish. Each player copies the format shown.

1. On your turn, choose one of the dice and roll it.
2. Write the number rolled in one of the squares. No erasures are allowed.
3. When all numbers are filled in, each player multiplies. The winner is the player with the largest product (as long as 0's are in the right places).
4. Play the game a few times.

Write about strategies that help you to win this game. How do you decide which die to pick? How do you decide where to write a number?

In Preparation

To help emphasize place value concepts and estimation skills, allow students to use calculators to perform the actual calculations.

Afterwards

Thinking Out Loud

As part of the discussion, use prompts such as these:

- How do you decide which die to choose?
- In what sort of situation would you choose the 4-faced die? the 10-faced one?
- How do you decide where to put a large number? a small number?
- Did you make the best possible placements?
- How would the game be different if the winner had the *least* product, not the *greatest?*

About Solutions

For the multiplication format shown, to make the greatest possible product from the numbers N_1, N_2, N_3, N_4 and N_5 (in order of smallest to greatest), those numbers should be placed in the positions shown. This placement can be explained in terms of which numbers are being multiplied. For example, $N_4 \times N_5$ is in fact multiplied by 100×10 in the final product, whereas the product $N_1 \times N_2$ represents only ones.

Dice

This game can be played with a regular die, but having the choice of the three different dice adds an interesting challenge. If a player requires a greater digit, it is best to pick the 10-faced die because the average roll is greater, unless getting a 0 would be too dangerous (for example, if the only space left were in the denominator of a fraction or in the first digit of a multi-digit number).

$$N_5 \quad N_3 \quad N_1$$
$$\times \quad N_4 \quad N_2$$

Going Further

You can assess students' understanding of this activity by showing a format of the game and explaining that you know in advance that you will roll the numbers 6, 4, 3, 1, and 1. Ask students how they would put the numbers in the format to yield the greatest result and the least result. Have them give reasons for their answers.

Students can design many variations on this game. (For instance, highest or lowest score can win.) Students can also design other game formats to use computational skills that they have been learning. Have students try to figure out strategies that help them to win in these games. Here are some samples:

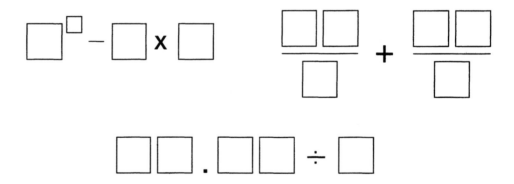

Any Order

Students roll four standard dice and use their colors as a code for which operations to use with them. The four operations can be done in any order: Students choose the order that will give them the greatest possible score. This activity leads students to realize that the order in which one does arithmetic operations makes a difference in the results. This, in turn, helps students to see the need for rules that designate the order of operations, and to realize that the most basic calculators don't "know" the rules.

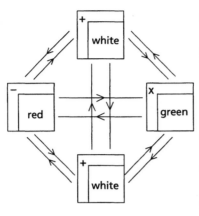

Any Order

This is a game for two to four players. You will need a gameboard like the one shown and four cube dice: one red, two white, and one green. Use a calculator if you wish.

1. Each player starts with 10 points.

2. On your turn, toss all four dice and place them in the squares by color.

3. Decide in which order to use the dice to compute the highest total possible. Start with the points you have already accumulated, and do the operations indicated.

   The winner is the player who has the highest score after four turns.

4. Discuss your results and compare how you chose the order in which you used the dice.

5. Try the game several more times until you think you have found some winning strategies.

6. Write about how you should play to get the most points.

In Preparation

Before students begin, make sure that they understand how to interpret a certain order of dice. For example, if they roll a 2 and a 4 on the white dice, a 6 on the red die, and a 3 on the green die, they might choose the order white 2, green 3, red 6, and white 4. If students start with 10 chips, their computation could be represented like this:

$$10 \xrightarrow{+2} 12 \xrightarrow{\times 3} 36 \xrightarrow{-6} 30 \xrightarrow{+4} 34$$

Note that this is not what is usually meant by the expression $10 + 2 \times 3 - 6 + 4$. Usually, we apply the rule for order of operations and interpret this as $10 + (2 \times 3) - 6 + 4 = 14$. If, however, one simply keys in these symbols on a calculator, the answer will be 34. Encourage students to write their results in the arrow format above so that they remember that this game defines order as working from left to right, which is the same way that a calculator operates.

Afterwards

Thinking Out Loud

As part of the discussion, use prompts such as these:

- Can you get different results if you use the dice in different orders?
- How did you decide in which order to use the dice?
- What is the most that you could get on the first turn?
- What is the worst possible roll that you could get?

Dice

- What would happen if you didn't start with 10 points?
- How could you change the rules to make the game different?

About Solutions
To get the most points in this game, the appropriate strategy is to use the two white dice first (addition order doesn't matter here), then the green die (multiplication), and finally the red die (subtraction).

If one does not start with 10 points, it is possible to be forced into having a negative number of points. This can make the game more interesting, but the strategy doesn't change. It would change, however, if the aim were to get the least points or to get the final score as close to 0 as possible.

Going Further
Suggest that students design variations of this game and try them out. Students can vary the game by having the winner be the player with the lowest score or the score closest to a certain number. They also can change the meaning given to the colors of the dice. Ask students whether their strategies are always the same. In particular, if they are aiming for the least number of points, does it work to simply reverse the order they would use to get the greatest number of points?

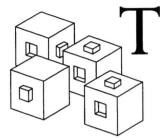

Connecting Centimeter Cubes

Thee connecting centimeter cubes, which come in 10 colors, can be joined together on four sides and in many configurations. Because the cubes fasten securely, constructions can be handled easily and examined from different sides. As a result, students have an opportunity to develop their spatial visualization skills and their ability to recognize three-dimensional patterns.

The cubes provide versatile models of metric units. Each edge is 1 centimeter long, each face has an area of 1 cm^2, and each cube has a volume of 1 cm^3 and a mass of 1 gram. Thus, the cubes are also ideal for providing the hands-on experiences that students need to develop an intuitive understanding of the metric system of measurement of length, area, volume, and mass.

Three connecting centimeter cube activities and teacher notes follow.

Building with Cubes

Students build a structure from a simplified floor plan and then draw the structure as it appears from the top, side, and front. Students then construct other buildings from top, side, and front views. Students also explore how to draw buildings on isometric grid paper.

This activity is designed to help students learn to imagine a three-dimensional object from two-dimensional views of it. This ability is essential not only for future work in mathematics but also for everyday situations in which students are called upon to interpret diagrams to operate appliances or instructions for assembling consumer products.

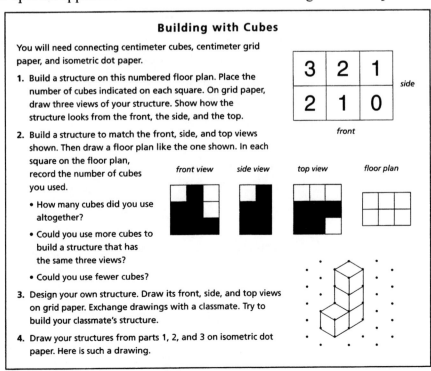

Building with Cubes

You will need connecting centimeter cubes, centimeter grid paper, and isometric dot paper.

1. Build a structure on this numbered floor plan. Place the number of cubes indicated on each square. On grid paper, draw three views of your structure. Show how the structure looks from the front, the side, and the top.

3	2	1
2	1	0

side *front*

2. Build a structure to match the front, side, and top views shown. Then draw a floor plan like the one shown. In each square on the floor plan, record the number of cubes you used.

front view *side view* *top view* *floor plan*

- How many cubes did you use altogether?
- Could you use more cubes to build a structure that has the same three views?
- Could you use fewer cubes?

3. Design your own structure. Draw its front, side, and top views on grid paper. Exchange drawings with a classmate. Try to build your classmate's structure.

4. Draw your structures from parts 1, 2, and 3 on isometric dot paper. Here is such a drawing.

Connecting Centimeter Cubes

In Preparation

Although some students already may have considered how two-dimensional images can describe three-dimensional objects, many most likely have had little experience in this area. Shadows are examples of two-dimensional images that give information about the object that throws the shadow. To provide an experience with a shadow, show how a block such as a wedge can cast at least the three different shadows shown in the illustration. The wedge can be placed on an overhead projector so that its shadows can be seen directly.

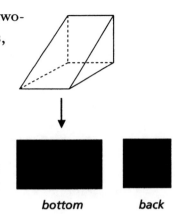

side   *bottom*   *back*

Since building a three-dimensional structure to match different views may well be new to students, you might want to model this type of activity before students begin. Show how you can construct a building that matches the required top view and then add to your structure to make it match the front view. After that, turn the structure and show that it doesn't match the side view and will need to be modified.

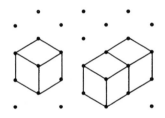

You might also give students some simpler drawing tasks on the isometric paper, such as drawing one cube or rods made of two or three cubes.

Afterwards

Thinking Out Loud

As part of the discussion, use prompts such as these:
- How did you start your construction?
- How did you decide where to put cubes?
- Does it matter from which side you look at the structure?
- Could you put in more cubes and still have the same front, side, and top views? Would using fewer cubes make a difference?

About Solutions

One approach is to use reasoning to place cubes to match the drawings, whereas another approach is to use trial and error, first placing cubes and then adjusting them to match the views. One strategy for building structures from the three views is as follows: Look at the top view and build a structure like that one cube tall; then look at the structure from the front and add cubes to make a tower as tall as indicated by the front view; finally, turn the structure to the side and take off whatever is extra. This will yield a structure with the greatest number of cubes possible. It may be possible, however, to remove some more cubes and still have the views stay the same.

Connecting Centimeter Cubes

For example, two solutions to part 2 of this activity are shown drawn on isometric dot paper.

Each dot on this paper is the same distance from the six closest dots, which is why it is called *isometric*, meaning "same measure."

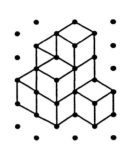

Going Further

Challenge students to think about how architects can describe the three-dimensional buildings they design on paper. Ask students to think of the school or some other nearby building. Can they draw the top, side, and front views? Have students try it and then go outside to check the accuracy of the drawings.

You can collect students' responses to parts 3 and 4 and use them to make a task file. On the front of a large file card, put the front, side, and top views of a student's structure. On the back, put the student's isometric drawing of one solution. Encourage other students to add more drawings of different solutions.

Patterns with Cubes

Students extend a pattern of three-dimensional structures by building and then visualizing the next ones. They go on to find the surface area and volume of these structures and look for numerical patterns.

This concrete experience with solid objects is essential to develop understanding of which aspects of a solid—surface area or volume—are being measured. If students first develop their own strategies and patterns for finding these measurements, their later study of standard formulas will be easier and more meaningful. Students initially describe the numerical patterns that they discover in different ways, perhaps using informal language, and, at a later stage, using algebraic notation and properties.

Patterns with Cubes

You will need connecting centimeter cubes.

1. Build the fourth structure in this sequence with connecting centimeter cubes. Then use words to describe what the fifth structure will look like.

First   *Second*   *Third*

(continued)

Connecting Centimeter Cubes

2. What happens to the length, volume, and surface area as the structure grows? Find and record these measurements in a chart like the one shown.

	First	Second	Third	Fourth	Fifth
Length of a side in cm (N)	3	4	5		
Volume in cm³ (V)	8				
Surface area in cm² (S)	32				

Make a list of patterns you noticed and strategies you used when completing this chart.

3. Repeat this activity with each set of structures shown.

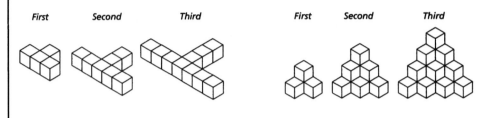

First   Second   Third   First   Second   Third

In Preparation

Students should be able to find area in square centimeters by counting squares. Make sure that they understand how they can find the surface area of a construction built with the connecting cubes by finding the sum of the areas of all faces showing. Students also should realize that the volume of such a structure is the number of cubes required to build it. Before students approach the activity, have all of them construct a simple form, possibly an L, as shown, and find its surface area and volume (26 cm²; 6 cm³).

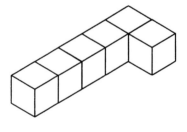

Afterwards

Thinking Out Loud

As part of the discussion, use prompts such as these:
- How did you decide which structure should come next?
- How did you change your structure to make the next one?
- What strategies did you use to find the surface area and volume of each structure?
- Can you explain why the patterns that you found exist?
- What other sequence of structures could you build?

Connecting Centimeter Cubes

About Solutions

The volume of each structure in part 1 can be found by simply counting cubes. The surface area can be found by looking at every structure from all of the six possible directions, each time counting square centimeters facing toward you.

Here is the chart of measurements that results from part 2.

	First	Second	Third	Fourth
Length of a side in cm (N)	3	4	5	6
Volume in cm³ (V)	8	12	16	20
Surface area in cm² (S)	32	48	64	80

The process of building the structures with cubes suggests some easier strategies for finding these surface areas and volumes. For example, to go from one construction to the next, rather than starting from scratch, remove two opposite corner cubes and insert four cubes, as shown. This makes it unnecessary to count the cubes in the new structure to find the volume; add four to the previous number, instead.

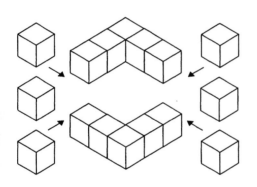

Adding these four cubes has the effect of adding four new units of surface area for each cube, or 16 in all.

There are other patterns in the chart. Perhaps the most obvious is that the surface area is four times the volume, which can be explained by noting that for each cube in one of the constructions, four faces are exposed. Symbolically, this is represented as $S = 4V$. The numbers for V are also all multiples of four—in fact, four times one less than the number of cubes on each side—which can be written as $V = 4(N - 1)$.

Another way to see a relationship is to think that to build a square with N cubes on a side, you need 4N cubes, but four of these are duplicated at the corners, leading to the formula $V = 4N - 4$.

These two formulas can be compared. In formal language, $4(N - 1) = 4N - 4$ by the distributive property of multiplication over subtraction.

The relationship between S and N can be extracted directly from the chart or explained as follows: Since $S = 4V$ and $V = 4N - 4$, it can be concluded that $S = 4(4N - 4)$ or $S = 16N - 16$ or, if you prefer, $S = 16(N - 1)$.

Part 3 yields the following two charts. In each, N is the length of the longest row of cubes.

	First	Second	Third	Fourth
N	3	5	7	9
V	4	7	10	13
S	18	30	42	54

Connecting Centimeter Cubes

	First	Second	Third	Fourth
N	2	3	4	5
V	4	10	20	35
S	18	36	60	90

In each chart there are many relationships to be discovered. In the first one, for example, S = 6N, S = 4V + 2, and V = 1/2 (3N − 1). (These relationships can be written in many different ways.) In the second pattern, each layer is the sum of consecutive whole numbers. For example, the third structure is made up of layers of 1, 1+2, 1+2+3, and 1+2+3+4 cubes. These numbers, 1, 3, 6, 10, ..., are the triangular numbers. The differences between volumes of structures in this pattern are the triangular numbers. The surface area is a multiple, six times a triangular number, which can be explained by the fact that from whatever direction one looks at one of these structures, one sees the same "staircase," a triangular number of cubes.

Going Further

For any one of these patterns, ask students how many more of the structures, of increasing size, they could build with 50 cubes. This will encourage students to see patterns rather than just to build and count.

Ask students to design another structure with cubes, make it grow, and see what happens to length, volume, and surface area.

You might point out to students that none of the examples given contain similar figures. One way in which structures can grow is by enlarging them in all dimensions so that structures remain similar. Have students build a small structure and then build another one similar in shape but twice as long, wide, and tall as the original. Then have them build a third structure that is three times as long, wide and tall as the original.

Ask students what they can say about how surface area and volume change if they enlarge figures in this way. Students will find that doubling the length of an edge results in multiplying surface area by 4 and volume by 8 and that tripling all dimensions results in multiplying surface area by 9 and volume by 27. In general, students will discover that enlarging a figure by a factor of N has the effect of multiplying surface area by N^2 and volume by N^3.

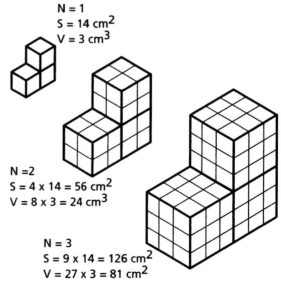

N = 1
S = 14 cm^2
V = 3 cm^3

N = 2
S = 4 x 14 = 56 cm^2
V = 8 x 3 = 24 cm^3

N = 3
S = 9 x 14 = 126 cm^2
V = 27 x 3 = 81 cm^2

Connecting Centimeter Cubes

The Price of Fruit

Students construct a spring scale with a rubber band and use the connecting centimeter cubes to calibrate the scale in grams. Then students use the spring scale to measure fruit, finding both the mass of a whole fruit and the inedible part. By collecting this data from more than one fruit, students are able to compare the cost per gram both of the whole fruits and of the edible portions.

After this direct experience of measuring in grams, students apply their results to solve problems connected with ratio. They also gain insight into the "nonexactness" of measurement; that is, that all measurement is approximate.

The Price of Fruit

You will need sticks of 10 connecting cubes, a paper cup, some paper clips, a rubber band, a ruler, some books, a strip of paper, some tape, and a calculator. You will also need some fruit (for which you know the price) and some plastic bags.

1. Construct a rubber-band scale. Look at the picture to see how to construct the scale.

   You will need the edge of a table or cabinet from which to hang the scale and a vertical surface on which to tape a strip of paper to calibrate, or label, your scale as follows:

   • Place a mark on the strip opposite the edge of the empty cup.
   • Put 10 cubes in the cup and mark the place where the edge is now.
   • Repeat this, adding 10 cubes at a time, until you have added all 200 cubes.
   • Each cube has mass of one gram, so you can use your scale to measure objects in grams. (You may want to change your set-up of the scale and link together two rubber bands to get a more sensitive scale.) If you want to calibrate your scale even further, find something that has mass of about 100 grams, and put it in the cup together with cubes.

2. Take a piece of fruit. Record how much it cost.
   • Find its mass.
   • Record the fruit's cost per gram.
   • Separate the edible part from the part that you throw away. Find the mass of each part. Record the cost per gram of the edible part.

3. Try this with a different fruit. Estimate whether it costs more or less per gram than the fruit you just measured. What about the cost per gram of the edible part?

4. For which fruit do you think there is the greatest difference between the cost of the entire fruit and the edible part? Is this reflected in the price of the fruit?

Connecting Centimeter Cubes

In Preparation

Before you show students the illustration of a spring scale, you might encourage them to brainstorm how they could construct a device for measuring the mass of small objects. Some students might suggest constructing a two-pan balance. In fact, if you have a two-pan balance available, students can use it instead of the spring scale in this activity. If students use the illustration as a model for their design, they will probably have to experiment with various rubber bands, ways to hold the device, and positions for fastening the calibrating strip.

Afterwards

Thinking Out Loud

As part of the discussion, use prompts such as these:
- Do you think your measurements are exact? Why?
- If you measured the same thing five times, would your scale give the same result each time? Try it.
- If someone else did this activity with the same apple, would they get the same result? Would they throw away the same amount?

About Solutions

Answers will depend on a lot of factors, such as how closely the fruit is peeled, how large it is, and its variety (for example, a juice orange may be different from a navel orange). If the same thickness of peel is removed from a small apple and a large apple, there will be a larger percentage of waste on the small apple. A more accurate picture is obtained if several fruits of each type are measured or if different groups of students (using different scales) measure the same things. Cost of fruit is usually determined by many factors other than the proportion of edible fruit, for example, if the fruit is in season, if it is grown locally or at a great distance, its popularity, and the type of store in which it is purchased.

Strictly speaking, a gram is a unit of mass (the quantity of matter) and not of weight (the result of weighing). Weight depends on the force of gravity, and objects of the same mass measured on planets that have different gravitational forces will have different weights. When weighing in a classroom, objects with the same mass will obviously have the same weight, so you don't have to be overly concerned if students' terminology is casual. Do make sure, however, that students understand that weight is not related to volume—that a given volume of lead does not weigh the same as the same volume of water.

Going Further

If several students collect data for this activity, suggest that they find a way to display their data on a graph (perhaps a double bar graph). Students might calculate and display the average percentage of each type of fruit that is wasted. Students could extend this investigation by considering how the price of a fruit is related to its popularity: They could collect data about favorite fruits of classmates and family and the prices of these fruits (per gram whole or per gram as eaten).

Suggest that students measure other common objects with their scales. Ask them to find the mass of a penny or a nickel in grams. This will be difficult to do if students measure only one penny. Instead they might try measure 10, 20, or more pennies. (The mass of a penny is about 3 grams, and the mass of a nickel is about 5 grams.)

If one of the fruits is approximately spherical, students might measure the diameter of the original fruit and also collect the peelings and lay them on centimeter grid paper to find the approximate surface area of the sphere. In this way they can verify (approximately) the formula for surface area S of a sphere with radius r, $S = 4\pi r^2$. Or they can cut the fruit in half, trace several copies of the circular face on paper, and find that the peel will fill about four of these circles.

Discuss with students other food items of which some is wasted and have them consider the price per amount eaten. For example, students could compare full broccoli heads versus already cut florets or boned chicken breasts versus whole ones. What does the consumer pay for convenience? What other factors might enter into our choices about how we buy particular foods?

Cuisenaire® Rods

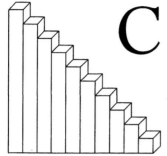

Cuisenaire Rods are a set of 10 different-colored rods, each with a square centimeter base. The shortest rod has a length of 1 centimeter and there is a rod of each multiple of that unit up to 10. Each length has its corresponding color, the shortest being white and the longest orange. Rich with visual and tactile appeal, Cuisenaire Rods can pose challenging problems for students of all ages.

For one thing, the rods can be used as geometric models. As such, they either can be used flat on a table to represent areas or in three dimensions, as solids with surface area and volume. The rods also can be used to model arithmetic: Addition is modeled by lining up rods end to end, then finding a rod, or rods, as long as the "train" formed. Because students can attach any numbers they want to the rods, the rods can easily provide models of fractions as well as whole numbers. For example, if the white rod is given a value of 1, the orange rod has a value of 10; if the white rod is 2, the orange rod is 20; if the orange rod is 1, the white is 1/10. This particular use of the rods helps students to visualize and generalize relationships among numbers, and thus gives real meaning to skills that are often learned at a purely symbolic level.

Three rod activities and teacher notes follow.

Perimeters

Students make several arrangements of the same four Cuisenaire Rods on centimeter grid paper and then investigate the perimeters of the shapes formed. Students discover that shapes with the same area can have different perimeters and that quite different shapes can have the same perimeter. This leads to the discovery of patterns relating to perimeter as well as varied strategies for finding it. By investigating perimeter in a nonformulaic setting, students are able to deepen and extend their understanding of this concept.

Perimeters

You will need Cuisenaire Rods and centimeter grid paper.

1. Take one red, two light-green, and one purple rod. Find at least 10 different ways to arrange them on the grid paper so that if you cut along their outlines, you would get a shape that stayed in one piece. Find the perimeter of each shape. Record each shape and its perimeter.

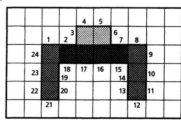

2. Answer these questions:

   • How many different perimeters did you find using these four rods? How could you check to see if you have all possible perimeters?

   • What is the least possible perimeter? Is there more than one shape with this perimeter?

   • What is the greatest possible perimeter? Is there more than one shape with this perimeter?

   • To find the perimeter, did you find more efficient ways than by counting every unit?

In Preparation

This activity requires prior experience with perimeter. If necessary, introduce or review this concept by drawing a shape on the overhead projector using the grid lines of a transparent centimeter grid. Tell students that in order to build a fence around this "field," they will need to buy fencing that comes in units the same length as the sides of the squares. Ask how many pieces of fencing will be required. Explain that this measurement is called the *perimeter.*

Contrast this problem with another in which students need to order turf that is sold in squares the same size as the squares on the grid. Ask how many pieces of turf will be required. Explain that this measurement is called the *area.* Have students experiment with drawing several "fields" on grid paper and then finding their areas and perimeters.

Afterwards

Thinking Out Loud

As part of the discussion, use prompts such as these:

- What strategies did you use to find perimeter?
- How can you change an arrangement of rods and still have the same perimeter? a greater perimeter?
- Do you think you found shapes with all possible perimenters? Why?
- Could a perimeter be an odd number? Why or why not?

About Solutions

Certain kinds of changes in the placement of the rods—for example, sliding a red rod 1 centimeter along one side of a purple rod—do not change perimeter. Other changes do. For example, rotating the red rod in the figure shown increases the perimeter by 2 centimeters.

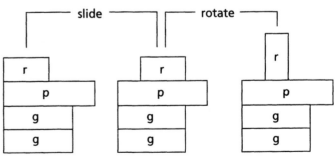

One way to find perimeter is by counting all of the units, but there are other strategies as well. For instance, the sum of the perimeters of all of the rods separately is 32 centimeters. When rods are placed side by side, this perimeter is decreased by twice the length of the contact. In the first two examples shown above, the contact length between the red and purple, purple and green, and green and green are, respectively, 2 units, 3 units, and 3 units, or 8 units in all. Thus, twice the total length of contact is 16 centimeters, and the perimeter of the shape is 32 − 16, or 16 centimeters.

Cuisenaire Rods

The perimeter is always an even number, a fact that can be explained by noting that the number subtracted from 32 (twice the length of contact) is even. Shapes can be found with perimeters equal to all even numbers between the minimum, 14, and the maximum, 26.

Going Further

Suggest that students use a different set of rods and see which of their discoveries remain true. You might also challenge students by asking them to explore how many different shapes they can find with the same perimeter but different areas.

Rod Sculptures

Students make different "buildings" from the same five Cuisenaire Rods and find their surface areas and volumes. Students discover that buildings of the same volume can have different surface areas and that buildings of very different shapes can have the same surface area.

Rod Sculptures

You will need a set of Cuisenaire Rods.

1. Take one light-green, two red, and two white rods. Put them together to make a building. Rods must touch on areas that are some multiple of square centimeters. Find the surface area and volume of your building.

2. Use another identical set of these five rods to make another building and then compare it with your original. Can you make
   - a building with the same volume but more surface area?
   - a building with the same volume but less surface area?
   - a building with different volume?

   What is the greatest and the least surface area of buildings that you can construct with these rods?

3. Write about your strategies for finding the surface area and volume.

In Preparation

In order to do this activity, students should know how to find surface area and volume of a construction. If necessary, select a purple rod. Show that it has a surface area of 18 square centimeters by "stamping" each of the six surfaces (sides) with a white rod. Show that the purple rod has a volume of four cubic centimeters by matching it to a train of four white rods.

Afterwards

*Thinking
Out Loud*

As part of the discussion, use prompts such as these:

- What strategies did you use to find the surface area?
- Can you predict when two buildings will have the same surface area?
  Explain how.
- Describe the buildings made with rods that had the most surface area; the least.
- Which was easier for you to find, the surface area or the volume? Why?

*About
Solutions*

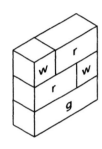

Surface area can be found by counting. The process is easier if the counting is organized in some way, such as by looking at the faces from each of the six possible directions. A different strategy, however, is similar to one described for perimeter in *Perimeters*. When the five rods are separate, the surface area is 46 cm². When two rods are put together, the total surface area is decreased by twice the area of contact. For example, to obtain the greatest possible surface area of a building made with five rods, each rod must touch the next on just 1 cm² of surface. One way to do this would be to arrange all the rods in one long rod, where the total area of contact is 4 cm²; the surface area is then $46 - (2 \times 4) = 38$ cm². The least possible surface area is obtained when there is the maximum contact. In the arrangement shown, the area of contact is 8 cm², and so the total surface area is $46 - (2 \times 8) = 30$ cm². Buildings can also be made with surface areas of 32, 34, and 36 cm².

*Going
Further*

Suggest that students try other combinations of rods. They might also compare this activity with *Patterns with Cubes*, the second activity under **Connecting Centimeter Cubes** (which also investigates surface area and volume). Students could in the same way invent sequences with Cuisenaire Rods and look for numerical patterns in the measures of length, surface area, and volume. Some examples follow:

Sequence 1

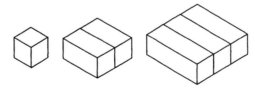

Sequence 3

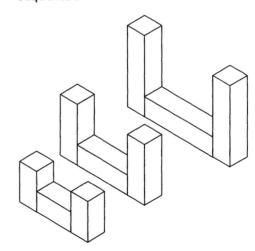

Sequence 2

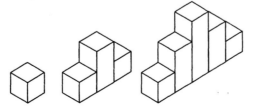

Cuisenaire Rods

Fraction Rod Race

Students play a game in which they use Cuisenaire Rods to model fractions. Rods are chosen by tossing dice; students add or subtract fractions by placing rods on a number line that they have constructed.

Students find that once a rod is assigned a value, certain fractions can be represented in terms of Cuisenaire Rods, that some fractions are equivalent, and that other fractions can't be represented by rods.

It is important for students to see physical interpretations of fraction addition and subtraction. Many think of addition of fractions entirely as symbolic manipulation and thus are unable to correct the common error of writing 1/2 + 1/4 = 2/6 by using physical intuition about what these symbols mean. A game such as *Fraction Rod Race* helps students to see that 2/6 is less than 1/2, and so the sum in this example must be wrong.

Fraction Rod Race

This is a game for two to four players. You will need a pair of regular dice and a set of 74 Cuisenaire Rods.

1. Each player should make a "race track" as follows. Draw a line segment the length of two dark-green rods. Label the points 0, 1 and 2 as shown.

2. Use other rods to find and label the points $\frac{1}{2}, 1\frac{1}{2}, \frac{1}{3}, \frac{2}{3}, 1\frac{1}{3}, 1\frac{2}{3}, \frac{1}{6}, \frac{5}{6}, 1\frac{1}{6}, 1\frac{5}{6}$.

3. Play the game. Take turns. On your turn, roll the dice and use the numbers shown to form a fraction. Use the smaller number as the numerator and the greater as the denominator. (If these numbers are the same, your fraction will be equivalent to 1.) If possible, take a rod that represents the fraction you roll and place it on your race track, starting at 0. If no such rod exists, your turn is over. On each turn, place the new rod next to the previous one to form a train. The object is to reach 2 exactly. If you roll a fraction that makes a train extending beyond 2, you must subtract a rod of that length from your train (exchanging rods if necessary).

4. Play the game at least two more times. Then write answers to these questions. From which positions do you have the best chance of winning on your next turn? From which positions is it impossible to win?

In Preparation

If students lack a solid conceptual background in fractions, first review how to find points corresponding to fractional parts on a number line. Cut enough strips of paper, all of the same length, for each student, and ask them to label the end points 0 and 1. Challenge students to find and label with fractions, as precisely as possible, as many of the points in between as they can. Students can then compare their markings and their strategies for making them. Making the "race track" in this activity will call on the same concepts but not exactly the same techniques.

Afterwards

*Thinking
Out Loud*

As part of the discussion, use prompts such as these:
- How did you use the rods to help you label the points with fractions?
- What lengths are you most likely to roll with the dice?
- What fractions can you roll with the dice that can't be shown with the rods?
- What is the smallest number of turns that it would take you to win? How likely is it that you will win in this many turns? Why?
- How could you write an addition or a subtraction sentence to express what you did on your last turn?

*About
Solutions*

Since the dark-green rod is the same length as six white rods, the rods can be used to represent any number of halves, thirds, and sixths. Some lengths can be rolled on the dice in several ways. For example, 1 can be rolled in six ways (1 and 1, 2 and 2, 3 and 3, 4 and 4, 5 and 5, and 6 and 6). Also, 1/2 can be rolled in six ways (1 and 2, 2 and 1, 2 and 4, 4 and 2, 3 and 6, and 6 and 3). There are 36 possible rolls of two dice (as you can see by drawing a chart like the one in the *About Solutions* section of *Two Dice or One?* under *Dice*). Thus, the probability of rolling 1 is 6/36, or 1/6, which is the same as the probability of rolling a 1/2. The probability of rolling any other length is less than this. The answer to the first question in part 4 is that you are most likely to win if you are either on 1 or on 1 1/2. If you are at any point less than 1, you can't win on the next turn.

When you overshoot the mark and have to subtract a rod, you can exchange your rods for whites to find your position on the number line. This is because the white rods represent sixths, and six is the common denominator for all fractions represented on students' number line.

Rolls of the dice that will not yield a fraction that can be represented by a rod are 1 and 4, 3 and 4, 1 and 5, 2 and 5, 3 and 5, and 4 and 5. Each of the above pairs can be rolled in two ways. Thus the probability of getting one of these pairs is 12 out of 36, or 1/3.

It is possible to win in only two turns if a fractional equivalent to 1 is rolled each time. The probability of rolling a 1 is 1/6, and the probability of this happening two times in a row is 1/6 x 1/6 = 1/36.

*Going
Further*

As seen in the section above, you can ask students many questions about probability in connection with this game. You can also ask how they could design dice so that they would be less likely to roll a fraction that can't be represented by rods in this game.

A further challenge would be to investigate how the game would change if students started with two brown rods (eight whites each) instead of two dark-green rods, or if they used 12 whites as the unit.

Two-Color Counters

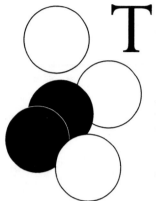

Two-Color Counters are plastic disks that are red on one side and yellow on the other. Pleasant to handle and easy to flip over, they are a perfect material to use for investigating probability. If, for example, students spill some onto a table, it is natural for them to first count yellows and reds and then to wonder about the chances of getting a certain number of each color on subsequent spills. Two-Color Counters can also be used to model whole numbers, fractions (a set of counters 1/3 of which are red), and integers.

Four Two-Color Counter activities and teacher notes follow.

Tossing Four Counters

In this activity, students gain experience in predicting, collecting, and interpreting data. As they toss four counters many times and record how many land red side up, students have the opportunity to discover that events that may at first appear to be equally likely, over the long run are not.

Tossing Four Counters

Work with a partner, taking turns tossing the counters, and record the results. You will need a graph like the one shown, four Two-Color Counters, and a container.

1. Put the counters in the container and shake it. Then spill the counters onto a flat surface. Count how many counters turn up red and record that number by putting an X on your graph. Do this once more and record your new results.

2. Predict which number will have the most X's if you spill the counters many times.

3. Now spill and record 30 more times.

4. When you are done, write your answers to these questions.

   • Did your prediction match your results?

   • Which number on the graph has the most X's?

   • Why do you think that happened?

   • How did your results compare to those of your classmates?

5. Do the Two-Color Counter activity *Arranging Four Counters*. Afterwards, compare your results from the two activities.

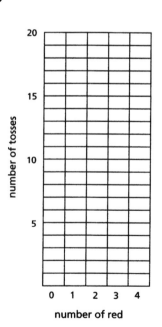

In Preparation

Before students approach this activity, ask in how many ways Two-Color Counters could land when tossed. You might also ask students how they could keep track of the results from tossing counters many times. This would give them a choice of recording methods before being introduced to the graphing used here.

Afterwards

Thinking Out Loud

As part of the discussion, use prompts such as these:

- Why did you predict the number you did?
- Did you get a particular number more or fewer times than other numbers? Why?
- Did anyone else's graph look just like yours?

About Solutions

One might initially assume that all possibilities—4 reds, 3 reds, 2 reds, 1 red, or no reds—are equally likely to happen. However, the results of tossing counters many times are almost certain to contradict this assumption. In general, after many tosses one finds that two *events*—no reds and 4 reds—occur fewer times than the other numbers, and that 2 reds occur most often. This can be explained by examining the *sample space*, which contains 16 different, equally likely *outcomes* with respect to color. Only one of these outcomes yields 0 reds, while 6 of them yield 2 reds. These outcomes are enumerated in the *Arranging Four Counters* activity.

To explore the notion that empirical data gets closer and closer to theoretical expectations as the number of trials gets larger and larger, the results of an individual trial of this activity can be compared to the combined results of several trials, as shown below.

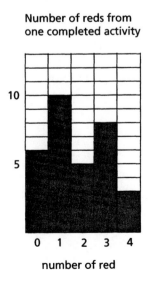

Number of reds from one completed activity

number of red

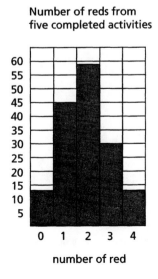

Number of reds from five completed activities

number of red

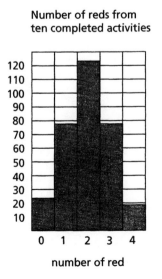

Number of reds from ten completed activities

number of red

Two-Color Counters

Note that relabeling the vertical scales on these graphs makes it easier to compare their shapes. When five trials were combined, for example, the scale was relabeled to be 5, 10, 15, … instead of 1, 2, 3, …. As more data is collected, the graphs become more symmetrical—that is, the number of 0's and 4's are about the same, as are the number of 1's and 3's. This can be explained by thinking of simply switching the names of the colors—if we call red yellow, and yellow red, then the number of times we would get no yellows in the old naming system would correspond to the number of times we would get 4 reds in the new system. (One presumes that there is no physical difference between reds and yellows.)

Going Further

After students have completed *Arranging Four Counters*, they might return to this activity and consider possible modifications. For instance, if a different number of counters were tossed, what would be the same and what would be different? You may want to point out that a computer is an ideal tool for gathering large amounts of data for these experiments. Perhaps you could work with a computer teacher in your school to find software that simulates this experiment, or even have students design such a simulation themselves.

Arranging Four Counters

This activity builds a theoretical probability model for the previous activity, *Tossing Four Counters*. Students investigate all the ways that four counters with two different colors showing can be arranged in a row. Students graph their results and compare them to those obtained from *Tossing Four Counters*.

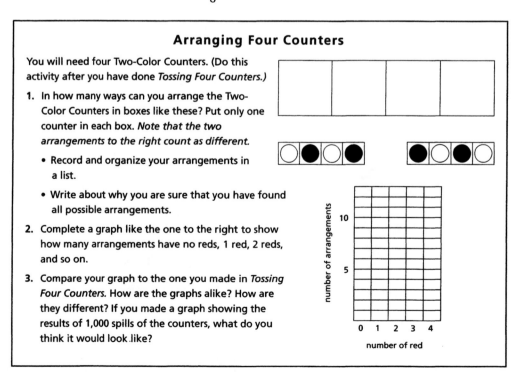

Arranging Four Counters

You will need four Two-Color Counters. (Do this activity after you have done *Tossing Four Counters*.)

1. In how many ways can you arrange the Two-Color Counters in boxes like these? Put only one counter in each box. *Note that the two arrangements to the right count as different.*

   • Record and organize your arrangements in a list.

   • Write about why you are sure that you have found all possible arrangements.

2. Complete a graph like the one to the right to show how many arrangements have no reds, 1 red, 2 reds, and so on.

3. Compare your graph to the one you made in *Tossing Four Counters*. How are the graphs alike? How are they different? If you made a graph showing the results of 1,000 spills of the counters, what do you think it would look like?

In Preparation

Have the class predict how many different arrangements there are using four Two-Color Counters. Model two different arrangements using the same amount of each color. For example, see the illustration at the right. Point out that although each contains 1 red and 3 yellow, the arrangements are considered different.

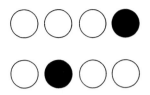

Afterwards

Thinking Out Loud

As part of the discussion, use prompts such as these:
- How did you record your results? Why?
- How many different arrangements did you find?
- Do you think you have them all? Why?
- Did you get the same number for arrangements with 1 red as for those with 3 reds? Why?

About Solutions

There are several "typical" systems for recording and organizing arrangements. One is to simply list the outcomes by using letters to represent the colors of the counters—for example, RRRR or RYRY. Breaking the possibilities into categories—for example, all arrangements with 0 reds, then 1 red, then 2 reds, and so forth, makes it easier to find all the solutions. The number of each category could be graphed as shown and compared to the graph of the empirical data collected in *Tossing Four Counters*.

A second way to organize arrangements is in a tree diagram, as shown in the middle of page 50, so called because the outcomes look like branches on a tree. This method corresponds to first looking for all arrangements with the first counter red or with the first counter yellow.

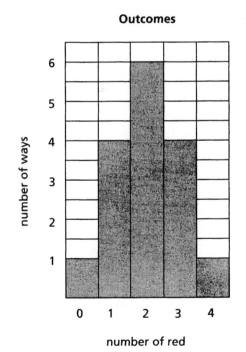

Outcomes

number of ways (vertical axis: 1, 2, 3, 4, 5, 6)

number of red (horizontal axis: 0, 1, 2, 3, 4)

Two-Color Counters

*Going
Further*

As an extension, ask students what they think would happen if they tossed 3 or 5 counters. Help students to organize all their results in a table, as shown, and look for patterns.

For example, adding the numbers in each row gives the totals of 2, 4, 8, 16, 32, ..., which are the powers of 2: 2^1, 2^2, 2^3, 2^4, 2^5, ... (This can be explained by looking at the tree diagram: With each new counter added, the number of branches in the tree diagram doubles.)

Students might also write the numbers in the table in a different form, as below, right. This arrangement, known as Pascal's Triangle, contains further patterns. Perhaps students will notice that each number is the sum of the two immediately above it. They might also notice that in the first few rows they can find the powers of 11: 11, $11^2 = 121$, $11^3 = 1331$....

Outcomes

numbers of red counters

	0	1	2	3	4	5
1 counter	1	1				
2 counters	1	2	1			
3 counters	1	3	3	1		
4 counters	1	4	6	4	1	
5 counters	1	5	10	10	5	1

Tree Diagram

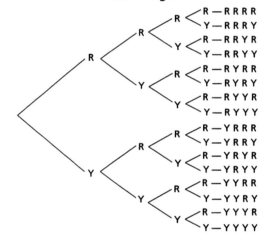

Pascal's Triangle

```
      1   1
    1   2   1
  1   3   3   1
1   4   6   4   1
1  5  10  10  5  1
```

Fraction Flips

Students use Two-Color Counters to solve riddles that focus on fractions. Some riddles have one solution, some have many, and some have none. Students hone their logical thinking skills while gaining experience with constructing models for fractions and finding equivalent fractions. The Two-Color Counters can easily be flipped, which makes them ideal for representing the clues in the riddles.

Fraction Flips

You will need Two-Color Counters.

1. Here are three riddles. One has no solution. Which one is it?

Riddle 1	Riddle 2	Riddle 3
I had some counters.	I had some counters.	I had some counters.
One third of them were yellow.	One third of them were yellow.	One third of them were yellow.
I flipped two of them over.	I flipped three of them over.	I flipped three of them over.
Then one half of them were yellow.	Then one half of them were yellow.	Then two thirds of them were red.
What counters did I start with?	What counters did I start with?	What counters did I start with?

2. Some riddles have many solutions. Consider this: I had some counters. One third of them were yellow. I flipped some of them over. Then one half of them were yellow. What can you say about the number of counters started with?

3. Using counters, make up your own fraction riddle for others to solve. Have someone try your riddle. Write it on an index card for a class set of riddles.

In Preparation

Students should have some experience with a discrete model for fractions. It is helpful for them to know, for example, that if there are 24 counters, of which 8 are red, then 1/3, 4/12, and 8/24 all describe the numbers that are red.

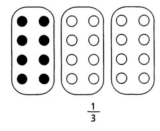

$\frac{1}{3}$

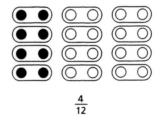

$\frac{4}{12}$

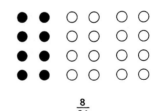

$\frac{8}{24}$

Afterwards

Thinking Out Loud

As part of the discussion, use prompts such as these:

- What strategies did you use to solve each riddle?
- How many counters could you start with in each riddle?
- What are some ways to have one third of the counters red? What do these ways have in common?
- Does it matter which counters you flip?
- Do you think that there might be other solutions?
- Why do you think there might be no solution?
- How did you go about creating your own riddle?

Two-Color Counters

About
Solutions

One strategy is to use trial and error—testing counters in multiples of 3 as suggested by the second clue in each riddle. Another is to reason as follows: In Riddle 1, when 2 counters were flipped, the fraction that was yellow changed from 1/3 to 1/2—a difference of 1/6. (Both counters must have been red because if there were 1 of each color, the situation would remain unchanged, and if both had been yellow, the fraction that was yellow would have decreased, not increased.) This means that 1/6 of the original number was 2, so there were 12 counters to begin with.

This same reasoning may be applied to Riddle 2: When 3 counters were flipped, the fraction that was yellow again changed from 1/3 to 1/2, a change of 1/6. If all counters flipped were red, then 1/6 of the original number was 3, and there were 18 counters to begin with. However, there is another solution to Riddle 2 because one need not assume that all 3 counters flipped were originally red: Perhaps 2 were yellow and 1 was red. This leads to the solution that originally there were 6 counters.

Riddle 3 has no solution because the situation is unchanged after the flip—1/3 yellow is the same as 2/3 red—but flipping 3 counters (or any odd number) must change the number of yellows.

The riddle in part 2 is more open than those in part 1. Two conditions vary—the number of counters to begin and the number of counters flipped. Here, all solutions must be multiples of 6.

Going
Further

Have students try to solve the riddles they created. Challenge students to make up riddles with 0, 1, 2, and more solutions.

Collecting Counters

In this game, students investigate probability as they decide what outcomes are possible, probable, and impossible. They collect red or yellow counters by following the directions on two spinners and adding or removing counters (based on the rule that a red and a yellow cancel each other out). Students score points according to the collection of counters they have at the end of each turn.

Since students' actions with the counters correspond to operations of arithmetic with integers, this activity serves as a springboard for learning about addition, subtraction, and multiplication of integers. In choosing which spinner to use first, students get the opportunity to investigate the effect of changing the order of operations on numbers.

Collecting Counters

This is a game for two to four players. You will need two transparent spinners and spinner backings and about 80 Two-Color Counters.

1. Each of you begins with 4 yellow counters.

2. On your turn, spin both spinners and follow these rules:
   - A red and a yellow cancel each other out.
   - You can add or remove an equal number of reds and yellows at any time. For example, suppose you have 2 reds and spin "Give away 2 yellows" and "Take 1 yellow." To do this, you can take 2 yellow-red pairs from the bank, return 2 yellows, take 1 yellow, and then return 1 yellow and 1 red.
   - At the end of each turn, you must have the fewest number of counters possible—all reds or all yellows.

3. Score points according to which counters you have at the end of your turn:
   1 point for more than 10 yellow counters;
   2 points for 0 counters;
   3 points for 5 counters (of either color).

4. The first player to score 10 points wins.

5. Play the game at least three times. Write about the game. What is the fewest number of rounds a player needs to win? Does it make a difference in which order you use the spinners? If so, give an example. Do you think that the way points are assigned in the game is fair? Did it affect how you played?

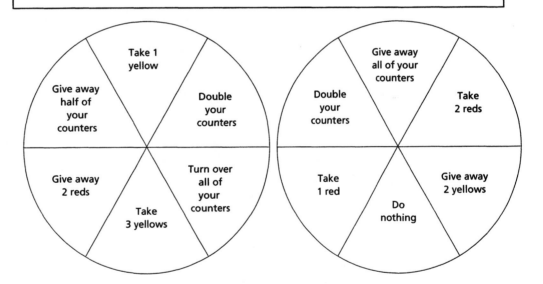

In Preparation

Students will need copies of the spinner backings on page 105. You might model the exchanges in this game by using transparent yellow and red chips on an overhead projector. Discuss what is meant by "Give away half of your counters." Students should understand that they should "round up" and give away the smallest number greater than one half. For example, if they have 5 counters, they should give away 3.

Two-Color Counters

Afterwards

Thinking
Out Loud

As part of the discussion, use prompts such as these:

- Why must you have only counters of one color at the end of a turn?
- Could you predict what you would have at the end of a turn before you actually took the counters? If so, give an example.
- How could you write the direction "Give away 2 reds" in another way?
- How would the game change if, before you spun, you could choose whether you wanted to use both spinners or just one?

About
Solutions

It is possible to win this game in four rounds, for example, by scoring 3 on three rounds and 1 on the fourth. If you start with 4 yellows, you can score 3 by ending up with 5 of the same color, using any of the nine combinations listed below:

> Take 3 yellows, Take 2 reds *or* Take 2 reds, Take 3 yellows;
>
> Take 3 yellows, Give away 2 yellows *or* Give away 2 yellows, Take 3 yellows;
>
> Give away 2 reds, Take 1 red *or* Take 1 red, Give away 2 reds;
>
> Take 1 yellow, Do nothing *or* Do nothing, Take 1 yellow;
>
> Turn over all of your counters, Take 1 red.

In the first two pairs of directions, "Take 2 reds" and "Give away 2 yellows" are essentially the same because these directions have the same effect. For all pairs of directions except the last, order doesn't matter. In general, order will make a difference when one of the directions is "Double your counters," "Turn over all of your counters," "Give away half of your counters," or "Give away all of your counters." Order won't make a difference if both directions involve giving or taking away a set number of counters.

Since there are 6 X 6 = 36 possible combinations of directions from the two spinners, the probability of getting 3 points on the first turn is $9/36 = 1/4$.

In the same way, you could find other probabilities—such as that of getting 3 points on the next turn if you have either 5 reds or 5 yellow counters—by listing which of the 36 combinations yield the desired results.

There are various strategies for playing the game. The probability of landing on "Give away all of your counters" on a turn is $1/6$, and whatever the other direction is, you will get 2 points if you use this direction second. On the other hand, you might aim to collect as many yellows as possible and try to get 1 point on each turn. The danger here is that you may land on "Turn over all of your counters."

Two-Color Counters

Going Further

Encourage students to explore what happens to the game if they change either the entries on the spinner, the way points are won, or the way the spinners are used. It is interesting to offer students the choice on each turn of using two spinners or one and, if one, which one it will be. On each turn, students will have to consider the probability of landing on certain directions in light of which counters they have.

In this game, if one assigns the values $^+1$ to the yellow counters and $^-1$ to the red counters and thinks of addition as joining, subtraction as taking away, and multiplication as repeated addition, one has a model for integers that is consistent with the usual rules for arithmetic. For example, $^+3 + {}^+2 = {}^+5$ and $^-2 + {}^-3 = {}^-5$ because when you join two sets of the same color, you end up with the same color. When you add counters of different colors, a certain number of pairs will cancel out (in fact the smaller of the two numbers), showing, for example, that $^+3 + {}^-5 = {}^+3 + {}^-3 + {}^-2 = {}^-2$. Taking away 2 reds has the same effect as adding 2 yellows, corresponding to the fact that $-(^-2) = {}^+2$. The direction "Turn over all of your counters" corresponds to the directions "Change the sign of the integer" or "Take the negative of the integer." This direction can be seen to correspond to multiplication by $^-1$. Multiplication by $^-2$ can be thought of as first multiplying by $^+2$ and then by $^-1$.

Once students are comfortable with the rules for manipulation of counters used in the game, make this connection to integers. Show how the game can be played in which turns are written as number sentences involving integers. Students might then design a new spinner with directions given in terms of integers. For example, they might specify "add $^+3$," "add $^-1$," "subtract $^+2$," "subtract $^-2$," "multiply by $^+2$," "multiply by $^-1$," and "multiply by 0."

Hundred Boards

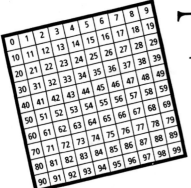

The hundred board is an arrangement of the numbers 0 to 99 in a square array. This arrangement provides a context for many investigations into properties of number systems and the development of algebraic language. It is also extremely useful for review of computational skills.

In the kit, you will find transparent color chips and a transparent hundred board. The chips, which come in five colors, fit nicely in the squares of the array while not obscuring the numbers. Visual patterns made by the color chips can bring numerical relationships to life. The transparent hundred board makes it easy for students to share their discoveries on an overhead projector.

Three hundred board activities and teacher notes follow.

Chip Patterns on a Hundred Board

Students are asked to find and describe numerical relationships that arise from placing the transparent color chips on the hundred board in a variety of patterns. Students discover that some of the relationships hold no matter where the chips are placed on the board; others don't. To explain how the numbers covered by chips are related, students are encouraged to describe one position in terms of another, for example, to call one chip X and to name the other chips in terms of X (perhaps X − 9 or X + 11). Hence, by working with concrete materials, students are encouraged to think in algebraic terms. At the same time, they have an interesting opportunity for computational practice.

Chip Patterns on a Hundred Board

You will need transparent color chips, a hundred board, and a calculator.

1. Put a blue chip somewhere on the hundred board, and surround it by 10 red chips, as shown. Then add the numbers covered by red chips. Repeat three more times.

   - Write about what you notice and why you think it happens.
   - Can you use an entirely different arrangement of 10 red chips and one blue chip and get the same result?

2. Arrange four red chips and four blue chips somewhere on the hundred board, as shown. Then add the numbers of each color. Repeat three more times.
   - Write about what you notice and why you think it happens.
   - Can you use a different arrangement of four red chips and four blue chips and get the same result?

(continued)

Hundred Boards

3. Take nine red chips and make a square on the hundred board, as shown. Position them so that the sum of the numbers covered will be each of the following:

162	360	378	596

- For each sum, write a description of your strategy or tell why you think that no such position exists.

In Preparation

This activity requires only basic arithmetic skills with whole numbers. Students should keep a calculator available, but encourage them to do as much mental arithmetic as possible.

Afterwards

Thinking Out Loud

As part of the discussion, use prompts such as these:
- Without looking at the board, how could you know what number is directly below another number? diagonally down to the right?
- How did you decide where to place your chips in a different arrangement to get the same result?
- In the square arrangement, how is the middle number related to the sum of all nine numbers?
- Describe how you found the sums for each example.

About Solutions

The pattern in part 1 is quite straightforward: The sum of the numbers covered by red chips will be 10 times the number covered by the blue chip. This can be explained by noting that pairs of red chips symmetrically arranged around the blue chip have a sum of twice the blue chip. It can also be explained with algebraic language when you call the blue chip X and name all of the others in terms of this, as shown.

Adding all of the expressions for numbers covered by red chips yields a sum of 10X. Although it may be easiest to use symmetry about a point when constructing other chip arrangements that give this result, nonsymmetric arrangements are possible.

In part 2, the sum of the numbers covered by each color will always be four times the number in the middle. This discovery leads to a strategy for part 3. The sum of

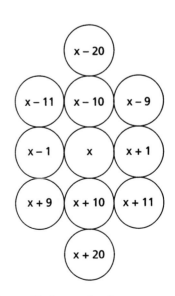

Hundred Boards

the numbers in the square of nine chips must be nine times the number in the middle. To find where the square of chips should be placed to get a given sum, simply divide the sum by 9 to get the number in the middle. For the numbers given, the results will be 18, 40, 42, and 66 2/9. Only 18 and 42 can be centers of squares on the hundred board. (Note that 40 is on the edge.)

Going Further

Students can come up with a variety of patterns similar to those presented in the activity. They might go beyond addition to explore multiplication and subtraction as well. You might suggest, for example, that students first arrange two red chips and two blue chips to be the corners of squares, as shown, and then find the difference between the products of chips of the same color. Can students find patterns? Can they explain them?

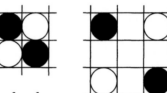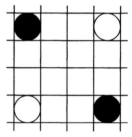

Encourage students to try this activity on arrays with a different amount of numbers in each row. Students will discover that on a calendar, for example, the number directly below the middle number X is X + 7, not X + 10, and the sum of the numbers above and below that middle number is (X − 7) + (X + 7), which is still 2X, just as it is on the hundred board.

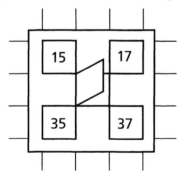

A dramatic way to display some of the patterns students discover is to make a screen out of construction paper like the one shown. Cut out the windows completely and leave the "door" attached on one side so that it can be opened. Put the screen over the hundred board on the overhead projector and have students add up the numbers they see and divide by four. Then open the door to see the result. Students may well wonder if the same thing happens when the screen is moved to other locations, and why.

Chips in a Row

Students play a game in which they choose four numbers by rolling four dice and then explore ways to combine these numbers (using any arithmetic process they like) to make numbers on the hundred board. The object of the game is to cover three numbers in a row with transparent color chips. This game has two educational purposes: to provide computational practice at the students' own level in a problem-solving context, and to encourage thinking strategies for placing chips on the grid to get three in a row. Although students should have calculators available, they will have to estimate and perform mental computation to know what to try on the calculator.

Chips in a Row

This is a game for two to five players. You will need transparent color chips, a hundred board, four standard dice, and a calculator. Each player uses a different color of chip. You may want to use a timer.

1. Warm-up

   If you roll four dice, you will get four digits. These digits can be used together with any arithmetic operations to make a number. For example, suppose you rolled 2, 4, 5, and 6. Here are some numbers that you could make:

   $$0 = \frac{(4 + 6)}{2} - 5 \qquad\qquad 16 = 25\% \text{ of } 64 \qquad\qquad 19 = 2.5 \times 6 + 4$$

   $$46 = 2 \times (4 \times 5) + 6 \qquad\qquad 80 = 24 + 56 \qquad\qquad 100 = (4 + 6) \times (2 \times 5)$$

   Find ways to make at least four more numbers.

2. Now, play the game.

   - Take turns rolling all four dice.
   - On your turn, choose a number you can make with the four digits you have rolled and write it down. Cover this number with a chip.
   - Do not put a chip on a number that has already been covered.
   - The first player to get three of his or her chips in a row—horizontally, vertically, or diagonally—wins the game.

   As a group, decide whether you want to use a timer to limit a player's time and, if so, how much time to give.

In Preparation

To enhance students' appreciation of the game and its possibilities, have them make as many numbers as they can from a set of digits such as the year or the number of their classroom. Invite students to come up with different solutions over the course of a few days and compare their strategies for forming new numbers.

Afterwards

*Thinking
Out Loud*

As part of the discussion, use prompts such as these:

- What playing strategies can you share?
- If someone has two chips next to each other, where should you place a chip?
- If someone has two chips in a row, can you stop that person from winning?
- What operations in arithmetic did you use? Could you use others?
- How did you decide how to get the number you wanted?
- How could you change the rules to make it *more difficult* to win? *easier* to win?
- Should there be a time limit in this game? Why or why not?

*About
Solutions*

When two players play this game, once a player has two in a row, the other player can't block both ends, so the first player always wins. This is not true if there are three or more players because players can work cooperatively to block another player. Whether or not players can take advantage of this strategy depends on their skill in finding expressions for numbers on which they want to place a chip.

There are many strategies for making a number with a particular roll of the dice. One is trial and error—experimenting with addition and subtraction to see what combinations can be made. Another is to work backwards, for example, by factoring the target number and seeing if factors can be made from available numbers. Still another is to explore all of the two-digit numbers that can be made from rolls of the dice. Skill at finding appropriate arithmetic expressions depends on good number sense and mental arithmetic, both of which develop with experience.

Going Further

Change the rules so that at the end of the game the player with the most rows of three (or four) wins. Have students compare this to the original rules. Is it easier? More difficult? To focus on particular computational skills, you can require their use in construction of the numbers. An example of this would be requiring that a fraction, a decimal, or a percent be used each time. You might also use a different set of dice, more or fewer than four, or perhaps some of the polyhedra dice. Finally, you might try using a different playing board. The illustration to the right suggests one possibility.

$\frac{1}{1}$	$\frac{1}{2}$	$\frac{1}{3}$	$\frac{1}{4}$	$\frac{1}{5}$	$\frac{1}{6}$	$\cdots$
$\frac{2}{1}$	$\frac{2}{2}$	$\frac{2}{3}$	$\frac{2}{4}$	$\frac{2}{5}$	$\frac{2}{6}$	$\cdots$
$\frac{3}{1}$	$\frac{3}{2}$	$\frac{3}{3}$	$\frac{3}{4}$	$\frac{3}{5}$	$\frac{3}{6}$	$\cdots$
$\frac{4}{1}$	$\frac{4}{2}$	$\frac{4}{3}$	$\frac{4}{4}$	$\frac{4}{5}$	$\frac{4}{6}$	$\cdots$
$\frac{5}{1}$	$\frac{5}{2}$	$\frac{5}{3}$	$\frac{5}{4}$	$\frac{5}{5}$	$\frac{5}{6}$	$\cdots$
$\frac{6}{1}$	$\frac{6}{2}$	$\frac{6}{3}$	$\frac{6}{4}$	$\frac{6}{5}$	$\frac{6}{6}$	$\cdots$
$\vdots$	$\vdots$	$\vdots$	$\vdots$	$\vdots$	$\vdots$	$\ddots$

Cover by Clues

Students use color chips to cover all numbers that satisfy certain properties. They explore how variations in these properties relate to visual patterns of the chips on the hundred board. Then, in a game format, students make decisions about which properties should be used to cover up the most numbers possible.

By manipulating the chips to satisfy these requirements, students develop a mental image of the hundred board that allows them to internalize and use the relationships among the numbers in our base-ten system.

Cover by Clues

This is a game for two to five players. You will need transparent color chips, a hundred board, and an envelope with clue cards. Each person uses a different color chip.

1. **Warm-up**

   Take a clue card out of the envelope. Each card describes some numbers on the hundred board. Each description contains a blank to be filled in with a number or phrase from a list. As a group, decide which number or phrase describes the most numbers and which choice describes the fewest numbers. Repeat this process for a different card. Then, put the cards back in the envelope.

2. **Now, play the game.**

   - Take turns drawing a card from the envelope. The person with the clue card that describes the fewest numbers covers those numbers with a chip. Put that clue aside. Return the other cards to the envelope and pass it clockwise to the next player.

   - On your turn, draw a clue card, decide how to fill in the blank, and use a chip to cover all of the still uncovered numbers that the card describes. Put that clue card aside.

   - Play until all the clue cards have been used. The player with the most chips on the board is the winner.

3. List the numbers that remain uncovered. Write as few clues as possible that will cause these numbers all to be covered.

4. Write about the visual patterns created by the clues.

In Preparation

Cut apart the clue cards that appear on page 106 and place the clues in an envelope. To do this activity, students should know the meaning of words such as *multiples*, *even*, *odd*, and *prime*. You may want to design additional clues that contain other appropriate vocabulary.

Afterwards

*Thinking
Out Loud*

As part of the discussion, use prompts such as these:
   - How do you know you have put chips on all the numbers for a particular clue?
   - How did you decide how to fill in the blanks in the clues?
   - How are patterns the same? How are they different?
   - Can you predict which clue would lead to a particular pattern?
   - Can you think of a clue that would cover exactly five numbers? exactly 50?

*About
Solutions*

A list of the clues appears on the next page.

Hundred Boards

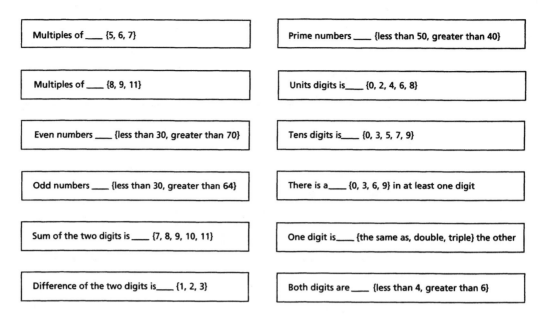

Multiples of ___ {5, 6, 7}

Prime numbers ___ {less than 50, greater than 40}

Multiples of ___ {8, 9, 11}

Units digits is ___ {0, 2, 4, 6, 8}

Even numbers ___ {less than 30, greater than 70}

Tens digits is ___ {0, 3, 5, 7, 9}

Odd numbers ___ {less than 30, greater than 64}

There is a ___ {0, 3, 6, 9} in at least one digit

Sum of the two digits is ___ {7, 8, 9, 10, 11}

One digit is ___ {the same as, double, triple} the other

Difference of the two digits is ___ {1, 2, 3}

Both digits are ___ {less than 4, greater than 6}

It is not always a good strategy to pick the replacement that covers the most numbers because, in the course of the game, many numbers will be already covered. A player must look at the situation on the board to decide which choice to make.

Many of the clues lead to striking visual patterns. Even numbers, odd numbers, multiples of 5, and numbers with the same units digit all lie in vertical columns. Numbers with the same tens digit lie in horizontal rows. Numbers with the same sum of digits, as well as multiples of 9, lie on diagonal lines rising from left to right. Numbers with the same difference of digits, as well as multiples of 11, lie on diagonals descending from left to right.

Going Further

An interesting challenge is to begin with an empty board and find the smallest possible combination of clues that will cover every number. Students might make up their own clues for this challenge. Stipulate that no single clue should cover more than twenty numbers to avoid trivial solutions such as "Cover all even numbers."

Encourage students to play the game on number boards with varying amounts of numbers in each row. They may discover that if N is the number in a row, then the multiples of any divisor of N make a vertical pattern, the multiples of any divisor of N − 1 make a diagonal pattern like /, and the multiples of any divisor of N+1 make a diagonal pattern like \.

Students can also extend this activity by creating the *Sieve of Eratosthenes* (named after a Greek mathematician in the third century B.C.) which is readily found in many standard mathematics textbooks. Using color chips to develop students' understanding of patterns of multiples makes this activity quick and yields information about factors. Doing this activity enriches and reinforces students' understanding of prime numbers.

AlphaShapes™

T he set of AlphaShapes consists of 52 translucent plastic pieces—26 green and 26 orange—that are designed to allow investigation of a number of geometric concepts. Each set contains two each of the following shapes: 6 different triangles, 11 different quadrilaterals, 3 different pentagons, 3 different hexagons, 1 circle, 1 oval, and 1 half circle. Every shape is identified with a letter of the alphabet and each angle is numbered. The set of 26 AlphaShapes of one color have a sufficient variety of properties—number of sides, measures of length and angle, parallelism of sides, symmetry, and convexity—for students to explore a wide range of relationships. Used together with an angle ruler (to measure length and angle), a hinged mirror (to explore symmetry), and a transparent centimeter grid (to measure area), AlphaShapes provide multiple experiences in developing definitions and in forming hypotheses and formulas related to symmetry and to measurement of length, angle, perimeter, and area.

Six AlphaShape activities and teacher notes follow.

Labels on Loops

This activity provides opportunities for reviewing and applying geometric language. Students compare the AlphaShapes by placing them in sets represented by two overlapping loops that are labeled with geometric properties. Such overlapping loops are commonly called a Venn diagram. First, students place the shapes in one of the four regions formed by the loops (note that outside both loops counts as a region). Once students have become familiar with this system of classification, they play "Guess the Labels." One player chooses labels for the loops in a Venn diagram but doesn't reveal their content. The other players try to guess the labels by placing shapes in the loops and being told whether or not the placement is correct. The goal is for students to guess the labels by trying as few shapes as possible.

Labels on Loops

You will need a set of labels, two loops of string, and 26 AlphaShapes of one color.
Work with a partner.

1. Warm-up
- Pick one label and find all of the shapes that have that property. Repeat this with another label.
- Make two overlapping loops of string, as shown. Pick two labels and put one on each loop. Place all of the AlphaShapes where they belong, inside or outside a loop or in the overlapping area.

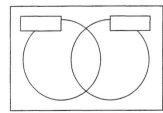

(continued)

AlphaShapes

2. Now, play the game.

- One person picks labels and places them face down on the loops.
- The other person guesses what is on the labels by placing a piece somewhere inside and waiting to hear whether the placement is correct. If the person who chose the labels says "yes," leave the piece in place. Otherwise, either move the piece or remove it and try putting down another piece.
- Record your guesses each time. The object is to guess the labels in as few tries as possible.
- Play at least four times; switch roles each time.

In Preparation

If students have had little experience with Venn diagrams as a visual means of communication, have them work with loops in a familiar context, perhaps using properties of numbers for the labels (for example, odd, multiples of 3, prime, two-digit).

For this activity, prepare labels for the loops such as those below, which are ready for reproduction on page 107. The collection of labels can be expanded to suit students' current geometric vocabulary. Have available a dictionary or math glossary to help students find the meaning of unknown terms. Provide some blank paper and scissors so that students can make up new labels of their own.

All sides are congruent.	All angles are congruent.	There is a right angle.
There are three angles.	There are four sides.	The shape has line symmetry.
The shape has rotational symmetry.	Exactly two sides are parallel.	Opposite sides are parallel.
No sides are parallel.	No sides are congruent.	Opposite sides are congruent.
At least two sides are congruent.		

Afterwards

Thinking Out Loud

As part of the discussion, use prompts such as these:

- What strategies did you use to select shapes to try in the game?
- Can you think of a label that applies to half of the AlphaShapes?
- Which pairs of labels make it possible to fill every region of the Venn diagram? Which make it impossible?
- Could you design some other shapes different from the AlphaShapes that would fit in this region?
- What other labels did you use in this activity?

<table>
<tr><td>
</td><td>The shapes in loop A all must have the property described on the loop label for A. Likewise, the shapes in loop B all must have the property indicated on the label for B. Shapes in the intersection of the loops, A ∩ B, have both properties, whereas the shapes that have neither property are outside both loops.</td></tr>
</table>

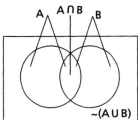

After experimenting with several sets of labels, two important conclusions can be derived: First, in order to guess what a label says, it is often as important to examine what is outside a loop as what is inside it. Second, when pairs of labels are contradictory, the intersection of the loops will be empty.

*Going
Further*

If students have not tried this themselves, have them arrange shapes in two loops labeled "All angles are congruent" and "All sides are congruent," respectively. Ask students to begin by considering just triangles. They will find that any triangle that has all sides congruent must also have all angles congruent. Next, have students consider quadrilaterals. Here, the loop labeled "All angles are congruent" will contain all rectangles, and the loop labeled "All sides are congruent" will contain all rhombuses; the only shapes that will be the overlap of these two loops are squares. This format for classification helps students to see that squares are a special kind of rectangle, that all squares are rectangles, and that all squares also are rhombuses.

As an interesting variation, students could invent label loops for arrangements of shapes like each of those shown, in which a shaded region means that no AlphaShape belongs in it and a white region indicates that at least one AlphaShape belongs inside.

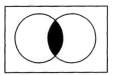  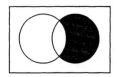 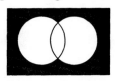

Encourage students to generalize about why certain labels make it impossible to have AlphaShapes in a particular region. Ask if there would still be no shapes whatsoever in that region, even if we included additional shapes not in the set of AlphaShapes.

Students also might explore similar activities using three loops instead of two.

Properties of Shapes

Students find all of the AlphaShapes that satisfy a given collection of properties and then determine which properties can be removed without requiring the removal of any of the shapes. Students then collect all the parallelograms in the AlphaShapes set, list as many properties as possible that they all satisfy, and repeat the process described in part 2 of removing superfluous properties.

AlphaShapes

Using manipulatives in this activity helps students to explore physically particular mathematical definitions, which leads to a deeper understanding of what mathematical definitions are. This activity shows that different definitions of a shape can be equivalent—that is, there can be two ways to describe the same set of shapes.

Properties of Shapes

You will need 26 AlphaShapes of one color.

1. Below are some properties of a shape. Find all the AlphaShapes that satisfy *all* of these statements.

   (a) There are four sides.

   (b) There are at least two congruent sides.

   (c) There are four angles.

   (d) The shape has at least one line of symmetry.

   (e) The shape has a right angle.

   (f) There are two pairs of congruent angles.

   (g) All angles are congruent.

   (h) There is at least one pair of parallel sides.

   (i) There are two pairs of parallel sides.

   (j) Both diagonals cut the shape into two congruent parts.

2. Cover up one of the properties above so that the same set of AlphaShapes you found for part 1 still satisfies the remaining properties. Then cover every other property that can be covered without making the set change. Make a list of the properties that remain.

   Uncover all of the properties and repeat this, but try to do it in a different way so that a different list of properties remains. Compare your lists with those that others have made.

3. Take AlphaShapes K, O, Q, U, W. List all the properties that are shared by *all* these shapes. Make sure that no other AlphaShapes satisfy all of the properties you have listed. Then repeat part 2, using this list.

In Preparation

Some experience with relationships among properties of geometric shapes would be helpful. You might play a guessing game in which students ask questions requiring "Yes/No" answers to determine which geometric shape you are thinking of. The goal of the game would be to guess the shape in twenty questions or less. In order to determine the shape, students need to put together information from various clues. They may come to realize that particular questions are unnecessary if their answers can be inferred from answers to previous questions.

Have available a dictionary or math glossary for students to use when they don't know the meaning of any geometric terms.

Afterwards

Thinking Out Loud

As part of the discussion, use prompts such as these:

- What strategies did you use to find all AlphaShapes that satisfy all of the properties?
- Why didn't a particular shape satisfy all of the properties?
- Would your answers be the same if, in addition to AlphaShapes, you could use *any* shapes that you could draw?

AlphaShapes

About Solutions

One approach to part 1 is to actually remove shapes that don't satisfy each property. There are 11 shapes that satisfy property (a), but of these, only 8 also satisfy property (b), and only 6 satisfy properties (a), (b), (c), and (d). Property (e) eliminates all but Q and W, which also satisfies properties (g)–(j). If additional shapes were considered, like the kite shown, Q and W, would satisfy properties (a) to (e) but, unlike Q and W, would not satisfy the entire list.

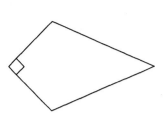

This example shows the difference between reasoning from a particular set of shapes, such as the AlphaShapes, and reasoning about all *possible* shapes.

There are many solutions to part 2: For instance, properties (a) and (g); properties (a), (e), and (i); properties (a), (e), and (j). Because any polygon always has the same number of sides as it has angles, properties (a) and (c) are interchangeable, making the following solutions possible: (c) and (g); (c), (e), and (i); (c), (e), and (j). For each of these solutions, if one statement is dropped, more shapes will satisfy the remaining list of statements.

The shapes given in part 3 all are parallelograms. As a set, they satisfy all of the properties in part 1 except (d), (e), and (g). Some of the additional properties that apply to all these shapes are: Opposite sides are congruent, opposite angles are congruent, the two diagonals bisect each other, and the sum of their angles is 360. The simplest minimal list is just properties (a) and (i).

Going Further

Repeat part 3 of the activity using another category of shape, such as a rhombus or a rectangle.

Discuss properties that are related to one another. For example, students might use a family-tree image to express their reasoning about properties of shapes. Students could write the properties on cards that can be affixed to the chalkboard. Arrows can be drawn between properties to indicate that "if I know this, then I know that," or "this implies that." Some examples are shown.

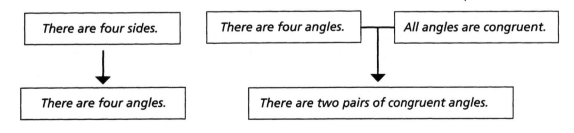

AlphaShapes

Ways to Find Area

Students investigate the areas of AlphaShapes by comparing each shape to the square and rectangle, whose areas in square centimeters are easy to find. Students then explore different techniques for finding area. Students also investigate the relationship between area and perimeter by discovering that shapes with the same area do not necessarily have the same perimeter. The design of the AlphaShape pieces allows students to discover various area relationships that are often taught primarily as formulas. Open-ended manipulative experiences such as this activity encourage students to reason about area relationships in ways that will later give meaning to the formulas.

Ways to Find Area

You will need 26 AlphaShapes of one color, a transparent centimeter grid, and an angle ruler.

1. Compare the area of each piece to pieces W and Q. Sort the shapes into five piles, as shown.

area less than W	area same as W	area greater than W but less than Q	area same as Q	area greater than Q

2. Write about your results and explain how you compared areas.

3. Find the area in square centimeters of at least two shapes from each of the five piles. If you can't find the area exactly, estimate. Write about how you found the areas.

4. Find some AlphaShapes with the same area. Measure their perimeters. What do you notice?

In Preparation

Students should have had some experience with finding area. The *Geoboards* activity *Patterns with Area* is related to this activity but uses geoboard units instead of square centimeters. Before students begin the activity, demonstrate the following methods for finding area, using the AlphaShapes and a transparent centimeter grid on an overhead projector.

- *Counting squares using the grid.* Put the transparent centimeter grid under W and count squares. If appropriate, use the formula for area of a rectangle, showing that it is a shortcut for counting squares (that is, multiplying the lengths of two sides gives 36 cm^2). Let students know that when using the grid on some shapes, they may need either to count fractional parts of squares or to estimate parts.

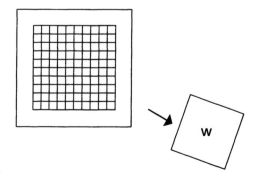

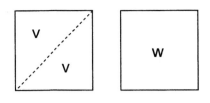

- **Comparing shapes.** Place 2 V's on the overhead projector to show that they form a shape identical to W. Explain that because two copies of V fit together to make W, V has half the area of W, or 18 cm².

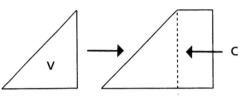

- **Combining shapes.** On the grid, put V over C so that only a rectangle is uncovered. The area of this rectangle will be found to be 12 cm²; thus, the area of C is 12 + 18 = 30 cm².

Afterwards

Thinking Out Loud

As part of the discussion, use prompts such as these:
- How did you decide what strategy to use to find the area of each shape?
- Which area measurements are rough estimates? Which are the most exact? Why?
- Can you find or draw shapes that have the same area and different perimeters?
- Can you find or draw shapes that have the same perimeter and different areas?
- Can you find or draw shapes that are not congruent that have the same area and the same perimeter?

About Solutions

Using one of the methods described above, the areas of 14 AlphaShape pieces can be found fairly exactly. Here is a summary of those areas and how they can be found.

Square W has area 36 cm², found either by counting or by using the formula *area of a rectangle = length × width*.

Rectangle Q has area 48 cm², found by the same methods.

Triangle V has area 18 cm²—half of W.

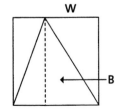

Triangle B also can be seen to have an area of 18 cm² because it can be cut into two triangles such that if copies of these triangles are joined to triangle B, the result is square W.

Triangles I and R both have area 24 cm²—both can be seen to be half of Q.

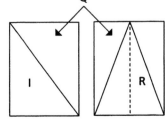

Parallelograms K and O both have area 48 cm²—each can be made from two copies of triangles R and I, respectively.

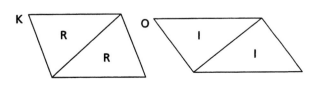

AlphaShapes

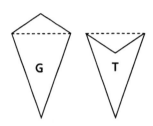

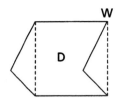

Triangle H has area 24 cm²—half of O.

Hexagon D has the same area as square W, 36 cm². One can make it by cutting a triangle out of one side of the square and pasting it back on the other.

Trapezoid C has area 18 + 12 = 30 cm², which can be seen by cutting it into two known parts, as described in *In Preparation*.

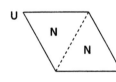

Kites G and T can be found by comparing them to shape R. One has a triangle with base 6 and height 2 added on to it; the other has the same triangle cut out from shape R. This triangle has area 6 cm²—it can be cut in half and reassembled to make a 2 × 3 rectangle. Thus, G and T have area 30 cm² and 18 cm², respectively.

Pentagon M can be seen as square W with two triangles cut out, both of which have base 6 and height 2 and hence, area 1/2(6 × 2) = 6 cm². Therefore, M has area 36 − 12 = 24 cm².

The areas of the other shapes can be estimated. For example, *Triangle N* has base 6 and height a little more than 5, so its area is about half that of a rectangle of area 30 cm², or about 15 cm². (The actual height can be found using the Pythagorean Theorem to be $3\sqrt{3}$, which is about 5.2 centimeters.) Using this estimate for shape N, some other areas also can be found.

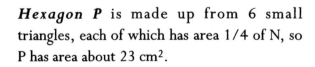

Rhombus U, which is twice N, has area about 2 × 15 = 30 cm².

Trapezoid S, made from three copies of N, has area about 3 × 15 = 45 cm².

Pentagon J, made from triangle N and square W, has area about 36 + 15 = 51 cm².

Hexagon P is made up from 6 small triangles, each of which has area 1/4 of N, so P has area about 23 cm².

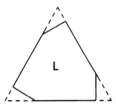

Hexagon L can be made from triangle N by cutting off three small triangles, each of which has area a little less than 1, so L has area about 15 − 3 = 12 cm².

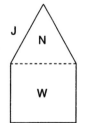

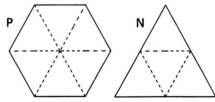

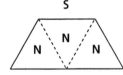

The areas of the remaining shapes can be estimated by direct counting, although students might use a pattern or formula to verify their estimates for the circle and semicircle (see *Circles*).

Circle X has area about 28 cm², **Semicircle Z** about 56 cm², and **Ellipse F** about 37 cm².

 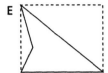

Quadrilateral A has area about 20 cm², and **Quadrilateral E** about 17 cm². Both can be seen to be almost half of a rectangle with a triangle cut off.

Pentagon F has area about 60 cm²—it is made up of 5 triangles with base 6 and height about 4.

Going Further

Students might explore certain sets of shapes having the same area; for example, quadrilaterals K, O, and Q or triangles H, I, and R. Investigations should lead to such generalizations as that quadrilaterals with the same base and height have the same area, or triangles with same base and height have the same area. Students might then try checking their generalizations by drawing on centimeter grid paper more quadrilaterals with the same base and height. Or you may want to use the informal methods for finding area to develop and give meaning to standard area formulas. For example, use the diagram at the right to show students why any parallelogram with base b and height h has the same area as the rectangle with the same dimensions.

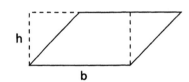

Tracing Angles

Students add together angles of an AlphaShape by tracing around each angle and placing one adjacent to the next. They then look for patterns by relating their results to the number of sides of each shape. Students discover that the sum of the angles of a triangle is a straight angle, of a quadrilateral a full circle, and so forth.

By doing this activity without the aid of a measuring device, students can develop physical intuition that will help them make sense of the standard formulas used by mathematicians. See *Ways to Measure Angles* of the **Reflect-It Hinged Mirror** for an approach to the angle sum of a polygon that uses measurement of angles in degrees.

AlphaShapes

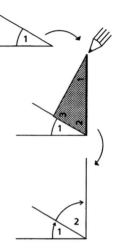

Tracing Angles

You will need all of the polygons found among 26 AlphaShapes of one color.

1. Take a triangle. The angles are numbered 1, 2, and 3.
   - Trace around angle 1 and label the angle.
   - Place the shape so that the vertex of angle 2 is on the vertex of the angle you traced and a side matches.
   - Trace around angle 2 and label it.
   - Repeat with angle 3.
2. Try this for at least two more triangles. Describe in writing what you notice.
3. Try the same thing with some four-sided shapes. Now the angles are numbered 1, 2, 3, and 4. Describe in writing what you notice.
4. Predict what will happen for all the rest of the AlphaShapes and explain your thinking. Then check your predictions.

In Preparation

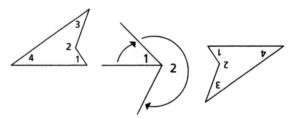

Students should have had some experience with the concept of an angle and with straight angles. To avoid difficulties with tracing angles of shapes that have more than four sides, model how to spiral tracings for shape J, as shown.

Notice that if you try shape E, angle 2 is greater than a straight angle, so angles 1 and 2 on your drawing should look like the one to the right.

Afterwards

Thinking Out Loud

As part of the discussion, use prompts such as these:
- Do you think your tracings are exact? Explain.
- How were your results for triangles the same? for quadrilaterals?
- Are your results true for all triangles (beyond those in the AlphaShape set)? Why?
- Can you describe your results in degrees? If so, how?

About Solutions

The triangular pieces B, H, I, N, R, and V each produce a straight angle. The quadrilateral pieces A, C, E, G, K, O, Q, S, T, U, and W each produce a complete circle, or 2 straight angles. Though more difficult to draw, pentagonal pieces D, J, and L create a circle and a half, or 3 straight angles, whereas the hexagonal pieces F, M, and P create two full circles for a total of 4 straight angles. To summarize, the sum of the angles of any polygon is always the number of straight angles equal to 2 less than the number of its sides. For example, the sum of the angles of all the 4-sided quadrilaterals equals 2 straight angles; the sum of the angles of a pentagon, which has 5 sides, equals 3 straight angles; the sum of the angles of a hexagon, which has 6 sides, equals 4 straight angles, and so on.

Since a straight angle measures $180°$, the above results can also be stated as follows: The sum of the angle measures of a triangle is $180°$; that of a quadrilateral is $360°$; a pentagon $540°$; and a hexagon $720°$. In general, the sum of the angle measures of any polygon with N sides is $180° (N - 2)$.

Going Further

To reinforce student findings, suggest that students measure each AlphaShape with the angle ruler and find the totals. Another way is to use coffee stirrers to divide each shape into triangles, using one vertex repeatedly, as shown. If students are convinced that the sum of angle measures of a triangle is $180°$, then they can count the number of triangles in each shape and multiply the result by 180 to find the sum of the angle measures.

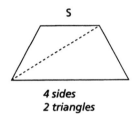

4 sides
2 triangles

M

5 sides
3 triangles

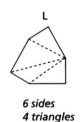

L

6 sides
4 triangles

Tessellations

Students use the AlphaShapes to try to make tiling patterns. They sort the shapes into those that can tessellate—that is, be used over and over again as tiles, leaving no spaces or overlaps—and those that can't. Students then look for patterns and form hypotheses about which kinds of shapes will tessellate and which kinds won't.

By physically using the pieces, students see dramatically just how properties of various shapes relate to whether or not those shapes can tessellate.

Tessellations

You will need a set of 52 AlphaShape pieces, some construction paper, and scissors.

Some of the AlphaShape pieces can be used to make tessellations—tiling patterns that completely cover a surface with no overlaps and no gaps. For example, S tessellates; F doesn't.

Tessellates   *Doesn't tessellate*

1. Separate all the AlphaShape pieces into two piles: shapes that can tessellate; shapes that can't tessellate. If you aren't sure where a shape belongs, cut out at least 10 construction paper copies and try to make a tessellation.

2. In what ways are the piles different? Explain in writing what makes only some shapes tessellate.

AlphaShapes

In Preparation

If students have never encountered tessellations before, point out how many there are in our environment—tiles laid on the floor, bricks on the wall, or other shapes that are arranged in decorative patterns. You might also demonstrate on an overhead projector how a tessellating pattern can be formed by repeatedly tracing around an AlphaShape. If students make tessellations from construction paper, their results will be easier to see if you provide more than one color of paper.

Afterwards

Thinking Out Loud

As part of the discussion, use prompts such as these:
- Which shapes did you know would tessellate without having to test them? Why?
- Did any of the tessellating patterns resemble each other?
- Do you think that triangles not included in this set would also tessellate? Why?
- Do all quadrilaterals tessellate? Why?
- Did you see other shapes in your tessellations?

About Solutions

For a tessellation to be possible, the space around a vertex must be filled. This means that the sum of the measures of the angles meeting at a point must equal 360°. All of the AlphaShape pieces tessellate except for F, L, M, X, Y, and Z. Since squares and rectangles tessellate, the half-squares and half-rectangles also tessellate. All

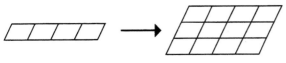

parallelograms tessellate, as you can see by arranging copies in a row to form strips and then putting these strips on top of each other. This tessellation can be easily drawn using two families of parallel lines.

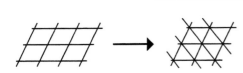

Every triangle is half of a parallelogram; hence, all triangles tessellate. (Add another family of parallel lines to the drawing.)

It is more surprising that all quadrilaterals tessellate. Here is a general procedure with which you can construct a tessellation from any quadrilateral. As an example, take quadrilateral A and trace around it. Mark the vertex of angle 1 with an X, as shown on the next page. Now rotate shape A, without turning it over, so that the vertex of angle 2 is at X and the side joining angles 1 and 2 matches the same side of the previously drawn shape. Repeat, but this time rotate so that the vertex of angle 3 is at X and, finally, so that the vertex of angle 4 is at X, each time matching the sides.

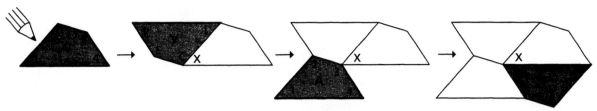

Note that the sum of the four angles of shape A is a full circle, so no gaps will be left. You can repeat this procedure around any other vertex. The same procedure works for any quadrilateral.

Some pentagons and hexagons will tessellate and some won't. For some polygons, you can reason that no tessellation is possible because there is no way to get a sum of 360° as the sum of the angle measures of the shape. For example, all the angles in pentagon F measure 108°, and 108 is not a divisor of 360, so the corners of shape F could not fill in a full circle with no gaps. You can reason in the same way to show that shapes L and M will not tessellate. On the other hand, shapes D, J, and P do tessellate, as shown.

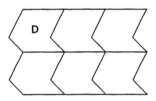
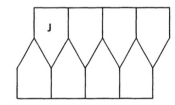
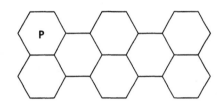

Finally, shapes X, Y, and Z don't tessellate because of their outwardly curved edges (which would need an inwardly curved edge to fit against).

Going Further

Once students have formed opinions about which AlphaShape pieces tessellate, challenge them to decide which of these statements they believe to be true, and why.

All squares tessellate.	*All rectangles tessellate.*
All rhombuses tessellate.	*All parallelograms tessellate.*
All trapezoids tessellate.	*All kites tessellate.*
All pentagons tessellate.	*All quadrilaterals tessellate.*
No pentagons tessellate.	*All hexagons tessellate.*
All convex quadrilaterals tessellate.	*No hexagons tessellate.*

Students should either be able to provide some sort of a reason why the shapes will tessellate or an example of a shape that won't tessellate.

AlphaShapes

Circles

Students measure the radius, diameter, circumference, and area of a variety of circles. They search for patterns and discover relationships among these measurements.

This direct involvement with measuring builds intuition about the units being used (centimeters and square centimeters) and brings to life relationships that sometimes are presented simply as abstract formulas. The dimensions of the circle and semi-circle in the AlphaShape set allow concrete investigation of the effects of doubling the diameter on the other measurements.

Circles

You will need at least five circles of different sizes (such as jar lids, coins, Two-Color Counters), 1 of AlphaShape X, 2 of AlphaShape Z's, string, an angle ruler, a compass, and the transparent centimeter grid.

1. Arrange all your circles, including the one made from the two Z's, in size order.
   - Estimate in centimeters, then measure the radius, diameter, and circumference of each circle. Record your findings.
   - Look for patterns and make a list of what you notice.

2. Draw or find a circle whose diameter is half as big as circle X. Use this circle along with circle X and the circle made from the two Z's to find the following information. Record your answers.
   - How do the measures of the radii compare?
   - How do the areas compare?
   - How does the measure of each circle's radius compare to its area?

3. Choose any of your circles. Make two new circles—one whose radius is twice as big and one whose radius is half as big as the circle you choose.
   - Predict the area of each circle.
   - Measure the area of each circle.
   - Explain, in writing, why your findings make sense.

In Preparation

Before beginning this activity, have students collect circular objects such as jar lids, paper cups, paper plates, and Frisbees™. Show students two ways to measure the circumference of a physical model of a circle—by stretching string around it and then putting the string next to a ruler, or by marking a point on the circumference of the circle and then rolling the circle along a ruler, starting with the marked point touching "0," and continuing until the marked point again touches the ruler.

Afterwards

Thinking
Out Loud

As part of the discussion, use prompts such as these:
- What patterns did you notice among your measurements?
- What would happen if you graphed your results?

- What would happen if you doubled the diameter of a circle? the radius? the circumference? the area?
- If you were to measure a new circle with radius 5, what would you predict to be the measurements of its diameter, circumference, and area?
- Would your patterns also work for AlphaShape Y?

About
Solutions

The following relationships should be discovered:

- The radius is half of the diameter or the diameter is twice the radius.
- The circumference is about three times the diameter and about six times the radius.
- If the diameter is doubled, the radius and the circumference are doubled, but the area is multiplied by 4. (In general, if the diameter is enlarged by any factor, the radius and the circumference are also multiplied by the factor, but the area is multiplied by the factor squared.)
- The area of a circle is about three times its radius squared. (If a square is drawn around the circle, it can be shown that the area of the circle is about 3/4 of the area of the square.)

Going
Further

Have students continue to investigate the effect of enlarging the diameter by investigating circles whose diameters are tripled. Ask students to hypothesize about what would happen to the circumference and the area if the diameter of any of their circles were tripled. You might even suggest using the transparent chips in this kit, since they are about 2 cm in diameter.

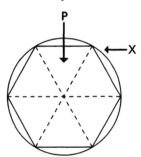

Students can be shown why the circumference and diameter of a circle are in the relation they are by looking at AlphaShape P, a regular hexagon, in the context of this activity. The perimeter of P is exactly three times its diagonal. Placed on top of Circle X, the diagonal of P is of the same length as the diameter of X, but the perimeter of P is a little less than the circumference of X.

Challenge students to draw a regular hexagon inside which the circle just fits. The perimeter of this hexagon will be little more than the circumference of the circle. Ask students to find that perimeter. (This hexagon should have sides measuring $2\sqrt{3}$ cm, or about 3.5 cm, so its perimeter is 6 × 2 × 3, or 2 × 3 × the diameter of the circle.)

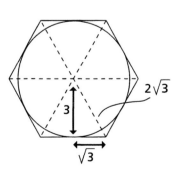

This activity shows that the circumference of a circle is between 3 and 3.5 times its diameter.

AlphaShapes

Students can also discover why the area of a circle is about three times the square of the radius by cutting the circle into pieces and rearranging these pieces into a parallelogram-like shape. In fact, the area is the product of the radius, r, and one half the circumference (1/2 × 2πr). To show this, have students use AlphaShape Z to trace a circle. Then have them cut out this circle, and fold it in half again and again to get 8 congruent, pie-shaped pieces.

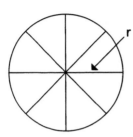

Then have student cut these pieces apart and rearrange them to make a shape that resembles a parallelogram except that the top and bottom are bumpy.

Therefore, the area of this figure—the product of its height and width—is r × 1/2 × 2πr or r, which is approximately 3 times the square of the radius.

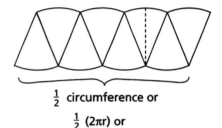

$\frac{1}{2}$ circumference or

$\frac{1}{2}$ (2πr) or

πr

Reflect-It ™ *Hinged Mirror*

T he Reflect-It Hinged Mirror consists of two 5 by 5 hinged mylar mirrors and a clear plastic base shaped like a protractor. The hinged mirror can be used with or without the base. Alone, it can be adjusted to any angle up to 180°. With the stand, the mirror's opening can be set and fixed at any of 10 angles.

In order to create patterns of reflections, either materials are placed between the faces of the mirror, or the hinged mirror is placed on top of any image. As the angle between the faces changes, the patterns change as well.

The hinged mirror captures the visual magic of a kaleidoscope while also providing a context for a variety of geometric investigations. The mirror can be used as a single mirror set at 180°, allowing investigation of line symmetry.

Three Reflect-It Hinged Mirror activities and teacher notes follow.

Ways to Measure Angles

Students use AlphaShapes™ to investigate the sum of the angle measures of polygons having varied numbers of sides. In the process, they use two strategies for measuring angles with just the hinged mirror: seeing how many copies of the angle fill up 360° and adding or subtracting two known angles. This activity helps to develop students' ability to estimate angle measures.

Students may also learn that the angle sum of any polygon is always the product of 180 and two less than its number of sides. Compare this approach with the *AlphaShapes* activity *Tracing Angles*, in which students discover this relationship without actually measuring angles. Students should experience and compare both approaches.

Ways to Measure Angles

You will need a hinged mirror and 26 different AlphaShapes.

1. Pick a triangle, a quadrilateral, a pentagon, and a hexagon from the set of AlphaShapes. If you added up the measures of all of the angles of each shape, which would have the greatest sum? the least? Why do you think so?

2. Use one of the following methods to check your estimates. Record your findings.

   • Put AlphaShape V—a triangle—between the mirror faces so that two of its sides lie flush with the faces. How many copies of the shape can you see? A full circle contains 360°. Divide 360 by the V's that you see, including the original piece. This is the measure of the angle between the mirrors. Repeat for the other angles of V.

   • Put together some angles that you know. For example, the measure of any angle of shape W, a square, is 90°. The smaller angles (1 and 3) of triangle V measure 45°. (Placing angle 1 or angle 3 of V on top of any angle of the square shows this 1:2 relationship.) Now put a square and the smaller angle of triangle V on top of angle 2 of shape C. They just cover the angle. This means that this angle measures 90° + 45° = 135°.

3. Now pick a different set of four AlphaShapes. Include a triangle, a quadrilateral, a pentagon, and a hexagon.

   • Estimate and/or measure the angles and find the sum for each shape.

   • Compare your results to those of the first set.

   • Write about what you notice.

Reflect-It Hinged Mirror

In Preparation

Before introducing the hinged mirror, encourage students to see how they can estimate and compare angles just by manipulating the AlphaShapes. When students first see the hinged mirror, let them explore freely with it. Then challenge students to find all the different angles they can measure with this mirror. Have them figure out the relationship between the letters on the base and measurement in degrees.

Afterwards

Thinking Out Loud

As part of the discussion, use prompts such as these:

- Which angles were easiest to measure? Why?
- Can you measure an angle greater than 180° with the hinged mirror? Why or why not?
- What strategies can you use to estimate an angle?

About Solutions

The hinged mirror can be used to measure angles whose measures divide 360 evenly; that is, whose measures are 360 divided by N, where N is a whole number. When a shape is put between the faces of the mirror, with a vertex at the hinge and its sides flush against both faces of the mirror, the shape and its reflections (or parts of them) fill in a complete circle. If an exact number (say N) of copies of the shape are seen, one can conclude that the measure of the angle at the hinge is $(360 \div N)°$. Below is a list of the AlphaShapes that contain such angles:

$360 \div 3 = 120°$	AlphaShapes A, D, P, S, U
$360 \div 4 = 90°$	AlphaShapes I, J, L, Q, C, V, W
$360 \div 5 = 72°$	AlphaShape B
$360 \div 6 = 60°$	AlphaShapes J, N, S, U, V
$360 \div 7 = 51°$ (approximately)	AlphaShape A
$360 \div 8 = 45°$	AlphaShapes C, D, V
$360 \div 9 = 40°$	AlphaShape A
$360 \div 10 = 36°$	AlphaShape E
$360 \div 11 = 33°$ (approximately)	AlphaShape H
$360 \div 12 = 30°$	AlphaShape M
$360 \div 18 = 20°$	AlphaShape H

When the angle becomes small, it is difficult to count all the reflected copies of the shape. It may be easier to measure a 20° angle (AlphaShape H) by seeing that two copies fit on top of a 40° angle (AlphaShape A). Other angle measures can be found or estimated using sums and differences of these angles.

*Going
Further*

You might have students practice estimation of angles by playing a whole-class game. Display the AlphaShapes on an overhead projector. Call out an angle measure and give students a minute or so to select, by letter and number, an angle whose measure they think is close to this number. Then measure angles that students have selected, or have them describe how they know their measures. Students can keep score by recording a running total of the differences between the measure of the angle they selected and the target measure. Their aim is to keep their score as low as possible.

See the *Going Further* section in *Tracing Angles* under **AlphaShapes** for additional ideas that relate to angle measure of polygons.

Lines of Symmetry

Students use the hinged mirror on AlphaShapes to find all lines of symmetry of each shape. They look for patterns relating the possible numbers of lines of symmetry to the number of sides of a polygon. Using the shapes with the mirror builds intuitive understanding of symmetry and facilitates exploration of its properties. Students will find that they need to go beyond the set of AlphaShapes to complete their investigations and will have to design shapes to meet specific criteria.

Lines of Symmetry

You will need 26 different AlphaShapes and the hinged mirror. Set the mirror at 180° for this activity.

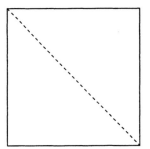

1. Place the mirror on the dotted line in this square. This dotted line is called a *line of symmetry*. A line of symmetry is a line along which you can place a mirror so that half of the original shape and its reflected image make the same image as the original shape. A square has four lines of symmetry. Find the other three.

2. Collect all the four-sided AlphaShapes. Determine how many lines of symmetry each one has. Agree or disagree with the following statements. Explain your reasoning.

   - A four-sided polygon has no more than four lines of symmetry.
   - A four-sided polygon can have any number of lines of symmetry up to four.
   - A four-sided polygon with four lines of symmetry has all of its sides and angles congruent.

3. How many lines of symmetry can a triangle have? a pentagon? a hexagon? Investigate and record your results. What patterns do you notice?

Reflect-It Hinged Mirror

In Preparation

To review the concept of symmetry, you might use the AlphaShapes on an overhead projector in a "Guess My Rule" activity. Put a pencil or straw down the middle of the screen and then, without telling students what the rule is, start sorting pieces—those having line symmetry on one side and those without line symmetry on the other. After placing a few pieces, ask students to keep their guess a secret but to test it as follows. Have students show their rule by placing additional pieces on the overhead. When most students seem to have guessed the rule, have one person verbalize it.

A common misconception is that any line that divides a shape into two congruent parts is a line of symmetry. Show that this is not the case by placing the hinged mirror on parallelograms K or O. Neither of these parallelograms has a line of symmetry, but their diagonals do divide them into two congruent triangles.

Afterwards

Thinking Out Loud

As part of the discussion, use prompts such as these:
- What examples of symmetry can you recognize in the environment around you without using the hinged mirror?
- To investigate the relation between number of sides and number of lines of symmetry of a polygon, did you find all of the shapes you needed among the AlphaShapes, or did you need to draw more shapes?
- How did you organize your data?

About Solutions

This chart shows the number of lines of symmetry for various shapes. *X* means no such shape is possible.

Number of lines of symmetry / Shape	0	1	2	3	4	5	6	More than 6
Triangle	B	V	X	N	X	X	X	X
Quadrilateral	A	Q S	R U	X	W	X	X	X
Pentagon	M	J	X	X	X	F	X	X
Hexagon	D	⌂	⬡	⬡	X	X	⬡	X
Other	⌣	Z	Y	⋔	✚	✦	✶	⊗

Some of the patterns related to symmetry that can be discovered in this activity are that:

- a polygon with N sides can have no more than N lines of symmetry;
- for each number of sides, one can construct a polygon with 0 lines of symmetry and one with 1 line of symmetry;
- for each number of sides, some numbers of lines of symmetry can't be realized; specifically, a polygon with N sides can't have exactly N − 1 lines of symmetry;
- a polygon with N sides and N lines of symmetry is a regular polygon—all sides and all angles are congruent;
- the number of lines of symmetry of a polygon with N sides must either be a divisor of N or 0. For example, a 4-sided polygon can have 0, 1, 2, or 4 lines of symmetry but not 3. A pentagon can have 0, 1, or 5 lines of symmetry, but not 2, 3, or 4.

Going Further

Suggest that students try to complete a chart like the one in *About Solutions* but do not include the sample shapes that have been put in each region. Display the chart for several days so that students can enter shapes as they find them. In some cases, students can refer to the AlphaShapes, and in others, they will need to design their own shapes.

Students might also investigate how line symmetry relates to rotational symmetry. They can use the AlphaShapes to test for rotational symmetry by tracing around one shape and then seeing if it can be rotated less than a full 360° turn (without turning it over) so that it again fits inside its outline. Students can sort the AlphaShapes into a Venn diagram with two loops labeled "Having Line Symmetry" and "Having Rotational Symmetry," respectively.

Finding Shapes

Students use the hinged mirror on a drawing of AlphaShape J to make various shapes. They sort a list of shapes into two piles—those that can be made using the hinged mirror on shape J and those that can't. When trying to make these shapes, students review the geometric vocabulary used in the description of the shapes. They also reason about how symmetry relates to both the way the mirror is used and to the properties of the shapes for which they are searching.

Reflect-It Hinged Mirror

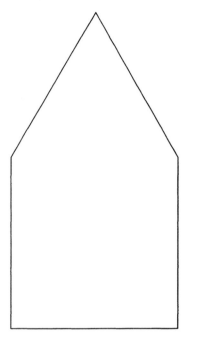

Finding Shapes

You will need a hinged mirror and 26 different AlphaShapes.

1. Put your hinged mirror on this drawing of shape J and move it around so that you see new shapes.

    • Find a way to make a square and a second rectangle that is not a square.

    • Use the mirror to make five more shapes and sketch each one.

2. Sort the following shapes into those that can be made by placing the hinged mirror on your figure and those that can't. Find a way to record the placement of the mirror so that you can discuss your results with someone else.

    > *rhombus (not a square)*
    > *parallelogram (not a rectangle or a rhombus)*
    > *trapezoid*
    > *quadrilateral with no sides congruent*
    > *kite (two pairs of congruent adjacent sides)*
    > *equilateral triangle*
    > *isosceles triangle (not equilateral)*
    > *scalene triangle*
    > *right triangle*
    > *pentagon with all of its angles congruent*
    > *pentagon with two right angles*
    > *hexagon*
    > *decagon*

3. Why are some shapes possible and others not? Give examples.

In Preparation

Students should know that the hinged mirror can be used in two ways in this activity—either by using the angle between the two faces or by using the mirror as one flat surface. (Students who have done *Ways to Measure Angles* and *Lines of Symmetry* will have experienced both uses of the hinged mirror.) Have a dictionary or math glossary available if students are unfamilar with some of the vocabulary in part 2.

Afterwards

Thinking Out Loud

As part of the discussion, use prompts such as these:

• How did you use the mirror to make each shape?

• Could you use the mirrors to make shapes that don't have line symmetry?

• Can you make each shape in a different way?

*About
Solutions*

Some shapes on the list can be made using either one face or both faces of the hinged mirror—for example, an equilateral triangle. Some shapes can only be made using both faces—for example, a pentagon with all of its angles congruent. Some shapes can be made using only one face—for example, the kite. Some can't be made at all—for example, the parallelogram, which is neither a rectangle nor a rhombus.

In general, shapes without line symmetry can't be made with a single mirror face. To find examples of the shapes on the list that have line symmetry, a strategy is to look for half of the shape in the figure. For example, to make a kite, try to form a nonisosceles triangle using the mirror edge and some part of the figure. This can be done by putting a single mirror on the dotted line shown and looking in the direction of the arrow. The other shapes on the list with line symmetry that can be made with a single face are a rhombus that is not a square, an equilateral triangle, a pentagon with two right angles, a hexagon, and a polygon with ten sides. Put the mirror on the dotted lines shown.

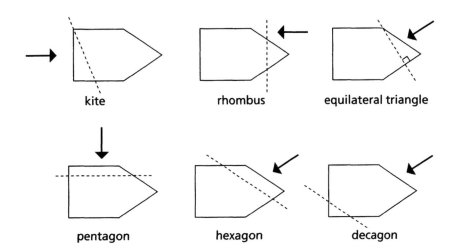

kite   rhombus   equilateral triangle

pentagon   hexagon   decagon

When you make shapes using both faces of the hinged mirror, you may find that you see different shapes when you look at the mirrors from different directions. These shapes need not have line symmetry. For example, if you put the hinged mirror

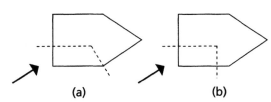

(a)   (b)

on the dotted lines shown in figure (a), you will see a trapezoid with no line symmetry if you look from the right direction. What you see, however, changes as you move.

Sometimes the shape made by the reflections seems to stay in the same place even if you move. For example, if you put the hinged mirror on the dotted lines shown in figure (b), you will see a rectangle made up of four smaller rectangles, and the image appears the same even if you shift your position.

Reflect-It Hinged Mirror

Shapes of this second type must have line symmetry and also rotational symmetry. The shapes on the list that can be made in this way are an equilateral triangle, a pentagon with all angles congruent, a hexagon, and a decagon.

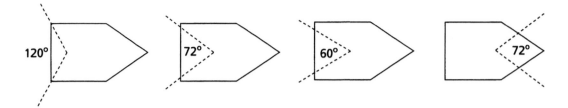

Going Further

Students can make up further activities of this sort by drawing a figure and then making a list of shapes to look for. You might challenge students to draw a figure different from shape J in which they could find as many of the shapes listed in the activity as possible.

Student Activities

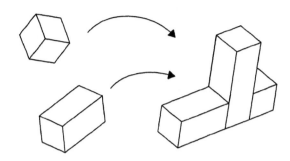

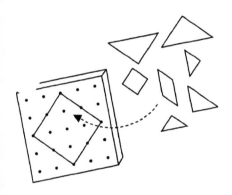

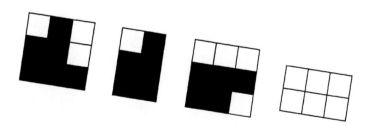

Mystery Shapes

You will need a geoboard, rubber bands, and sets of clue cards.

1. What is the mystery shape? Read all the clues, then use your geoboard to identify the shape.

> **Clue Set 1**
> It has four sides.
> All of its angles are congruent.
> One side is twice as long as another side.
> The rubber band does not touch the center peg.

What shape did you find? Can you find another shape to fit these clues? Record your solutions.

2. Find another mystery shape. Use another set of clue cards. Decide whether to work alone or with three classmates. If you work in a group, each person should

- have a geoboard and one clue.
- read his or her clue aloud.
- make a shape that satisfies each clue.
- touch only his or her own geoboard.

3. Repeat this activity. Use another set of clue cards.

4. Make a new mystery shape. Write four clues for it. Have others use the clues to make four different shapes. Add your clues to a class set.

Clue Set 2	**Clue Set 3**	**Clue Set 4**
It has a right triangle.	It has four sides.	It is not convex.
It has one interior peg.	It has no line of symmetry.	It has five sides.
It is isosceles.	Its area is six square units.	It has no interior pegs.
It uses a corner peg.	Just one pair of opposite sides is parallel.	It has a right angle.

Geoboards

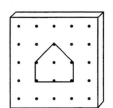
  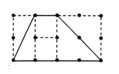
Squares on a Triangle

You will need a geoboard and geoboard dot paper.

1. Make as many squares of different sizes as you can on the geoboard. Record them on dot paper.

2. Find the area of each square.

3. Make this right triangle on your geoboard or on dot paper. Build squares on each side of the triangle and find their areas. Record the sum of the areas of the two smaller squares.

4. Repeat this procedure for the two triangles below.

What do you observe about the sum of the squares built on the two shorter sides of a right triangle and the area of the square built on the longer side? Write down an idea that might always apply.

Tangrams and the Geoboard

You will need a transparent geoboard, a rubber band, and one set of tangram pieces.

1. Make a square like the one to the right. Turn the geoboard over. Answer the following question by placing tangram pieces on top of the geoboard:

   What fractional part of the square is each of the tangram pieces?

   Trace each piece. Record the fraction inside and write a convincing argument explaining why your answer is correct.

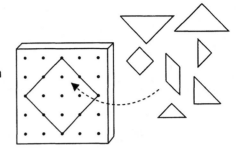

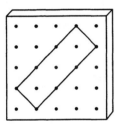

2. Make a rectangle like the one to the left. What fractional part of the rectangle is each of the five different pieces?

3. The small triangle has a different value when used in the square than it does when used in the rectangle. Why?

4. Make a shape of which the large triangle is 1/2. What fractional part of this shape is each of the other four pieces? Write about your strategies for answering this question and any patterns that you notice.

Similarity and Triangles

You will need a set of tangram pieces, an angle ruler, a transparent centimeter grid, and a calculator.

1. Take one of each of the three different sizes of triangles. Make a fourth triangle with the remaining pieces: large triangle, small triangle, square, and parallelogram. Arrange the four triangles in front of you.

2. Compare the triangles in as many ways as you can. (You can measure them with the angle ruler or the transparent centimeter grid.) Find how the triangles are the same and how they are different. Record your findings.

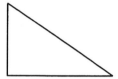

Mathematicians would say that the four triangles in front of you are similar to each other, but none of them is similar to the one on the right. Why do you think this is so?

3. Take the parallelogram, then use the other tangram pieces to make two more parallelograms, one using the two large triangles and the other

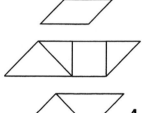

using the remaining pieces (the square, the two small triangles, and the medium triangle). Arrange the three parallelograms in front of you, as shown.

Mathematicians would say that the top and bottom parallelograms are similar to each other but neither is similar to the one in the middle. Why do you think this is so?

4. What *are* similar shapes? Write a definition to help someone understand.

Shopping for Tangram Pieces

You will need a set of tangram pieces.

Suppose the manufacturer of the four tangram sets must buy each color of dye at a different price. So, both color and size will determine the cost of each piece. For example, because the square can be made from two small triangles, the red square will cost twice as much as the small red triangle.

1. Use tangram pieces to figure out the cost of the pieces listed on the chart. Copy the chart and record your answers. List any patterns you notice.

	small triangle	medium triangle	large triangle	square	parallelogram
red	5¢			10¢	
green		12¢			
blue			16¢		
yellow					6¢

2. Fill in the outline with red and yellow pieces so that the shape will cost 56 cents. In writing, explain the process you went through to solve this problem.

3. Make your own tangram shape. Trace its outline. Figure out what your shape costs. Then tell someone the amount and have that person find the colors and pieces that will complete your outline and provide prices for each piece.

Describing Dice

You will need a partner and a set of polyhedra dice.

1. Pick two polyhedra dice. Make a list of ways in which the two dice are alike. Make another list of ways in which they are different. Try this again for another pair of polyhedra dice.

2. One way to describe dice is to count aspects of them. You might count the faces (the flat surfaces), the edges (the line segments where two faces meet), or the vertices (the points where edges meet). Make a table showing the numbers of faces, vertices, and edges of each die and describe any patterns that you find.

3. For each die, write a description that applies only to it and not to any other die. Can you do this in more than one way? For example, "the die that has 3 triangular faces meeting at a vertex" describes the tetrahedron die. Two other unique descriptions would be "the die with 3 numbers to each face" and "the die with 6 edges."

Rolling Threes

These are two games for two to six players. You will need a set of polyhedra dice. Share the dice so that each person is responsible for one to three of them.

Game 1   All of the dice are rolled at the same time, over and over. The winning die is the first to land on 3 three times.

- Predict which die you think will win.
- Play this game at least three times or until you feel sure about what the results will be any time you play. Keep a record of your results for each game.
- Discuss your results.
- Write about why your predictions did or did not match your results.

Game 2   This is the same as *Game 1* except that a die wins if it is the first to land on *multiples* of 3 three times.

- Predict which die you think will win.
- Write about how you made your prediction.
- Play this game as before.
- Compare your results to your prediction.

Two Dice or One?

You will need a partner and a set of polyhedra dice.

Suppose you will win $100 dollars if you roll a 3! You have a choice of which dice to roll:

- just the 10-faced die *or*
- both the 8-faced die and the cube die, where you take the sum of the numbers rolled.

Which would each of you choose? Explain why.

Now, you and your partner should each roll one of the two choices 20 times. Which of you got a 3 more often? Did this agree with your choice?

Compare your results with those of your classmates.

Making the Most of It

This is a game for two to six players. You will need only the polyhedra dice with single-digit numbers. Use a calculator if you wish. Each player copies the format shown.

1. On your turn, choose one of the dice and roll it.

2. Write the number rolled in one of the squares. No erasures are allowed.

3. When all numbers are filled in, each player multiplies.

   The winner is the player with the largest product (as long as 0's are in the right places).

4. Play the game a few times.

Write about strategies that help you to win this game. How do you decide which die to pick? How do you decide where to write a number?

Any Order

This is a game for two to four players. You will need a gameboard like the one shown and four cube dice: one red, two white, and one green. Use a calculator if you wish.

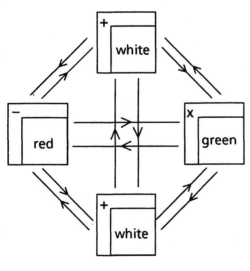

1. Each player starts with 10 points.

2. On your turn, toss all four dice and place them in the squares by color.

3. Decide in which order to use the dice to compute the highest total possible. Start with the points you have already accumulated, and do the operations indicated.

   The winner is the player who has the highest score after four turns.

4. Discuss your results and compare how you chose the order in which you used the dice.

5. Try the game several more times until you think you have found some winning strategies.

6. Write about how you should play to get the most points.

Connecting
Centimeter Cubes

Building with Cubes

You will need connecting centimeter cubes, centimeter grid paper, and isometric dot paper.

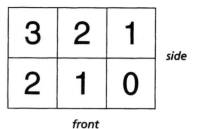

side

front

1. Build a structure on this numbered floor plan. Place the number of cubes indicated on each square. On grid paper, draw three views of your structure. Show how the structure looks from the front, the side, and the top.

2. Build a structure to match the front, side, and top views shown. Then draw a floor plan like the one shown. In each square on the floor plan, record the number of cubes you used.

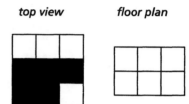

| *front view* | *side view* | *top view* | *floor plan* |

   • How many cubes did you use altogether?

   • Could you use more cubes to build a structure that has the same three views?

   • Could you use fewer cubes?

3. Design your own structure. Draw its front, side, and top views on grid paper. Exchange drawings with a classmate. Try to build your classmate's structure.

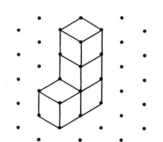

4. Draw your structures from parts 1, 2, and 3 on isometric dot paper. Here is such a drawing.

Patterns with Cubes

You will need connecting centimeter cubes.

1. Build the fourth structure in this sequence with connecting centimeter cubes. Then use words to describe what the fifth structure will look like.

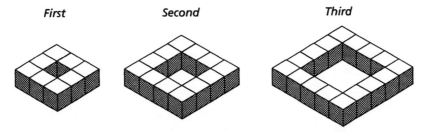

First   *Second*   *Third*

2. What happens to the length, volume, and surface area as the structure grows? Find and record these measurements in a chart like the one shown.

	First	*Second*	*Third*	*Fourth*	*Fifth*
Length of a side in cm (N)	3	4	5		
Volume in cm³ (V)	8				
Surface area in cm² (S)	32				

Make a list of patterns you noticed and strategies you used when completing this chart.

3. Repeat this activity with each set of structures shown.

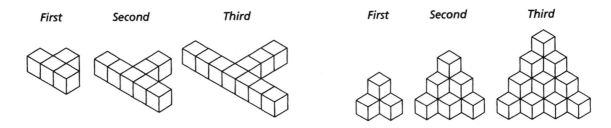

First   *Second*   *Third*   *First*   *Second*   *Third*

The Price of Fruit

You will need sticks of 10 connecting cubes, a paper cup, some paper clips, a rubber band, a ruler, some books, a strip of paper, some tape, and a calculator. You will also need some fruit (for which you know the price) and some plastic bags.

1. Construct a rubber-band scale. Look at the picture to see how to construct the scale.

   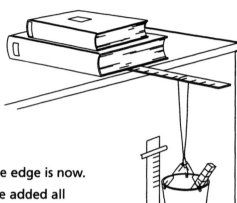

   You will need the edge of a table or cabinet from which to hang the scale and a vertical surface on which to tape a strip of paper to calibrate, or label, your scale as follows:

   - Place a mark on the strip opposite the edge of the empty cup.
   - Put 10 cubes in the cup and mark the place where the edge is now.
   - Repeat this, adding 10 cubes at a time, until you have added all 200 cubes.
   - Each cube has mass of one gram, so you can use your scale to measure objects in grams. (You may want to change your set-up of the scale and link together two rubber bands to get a more sensitive scale.) If you want to calibrate your scale even further, find something that has mass of about 100 grams, and put it in the cup together with cubes.

2. Take a piece of fruit. Record how much it cost.
   - Find its mass.
   - Record the fruit's cost per gram.
   - Separate the edible part from the part that you throw away. Find the mass of each part. Record the cost per gram of the edible part.

3. Try this with a different fruit. Estimate whether it costs more or less per gram than the fruit you just measured. What about the cost per gram of the edible part?

4. For which fruit do you think there is the greatest difference between the cost of the entire fruit and the edible part? Is this reflected in the price of the fruit?

Perimeters

You will need Cuisenaire Rods and centimeter grid paper.

1. Take one red, two light-green, and one purple rod. Find at least 10 different ways to arrange them on the grid paper so that if you cut along their outlines, you would get a shape that stayed in one piece. Find the perimeter of each shape. Record each shape and its perimeter.

2. Answer these questions:

   - How many different perimeters did you find using these four rods? How could you check to see if you have all possible perimeters?

   - What is the least possible perimeter? Is there more than one shape with this perimeter?

   - What is the greatest possible perimeter? Is there more than one shape with this perimeter?

   - To find the perimeter, did you find more efficient ways than by counting every unit?

Rod Sculptures

You will need a set of Cuisenaire Rods.

1. Take one light-green, two red, and two white rods. Put them together to make a building. Rods must touch on areas that are some multiple of square centimeters. Find the surface area and volume of your building.

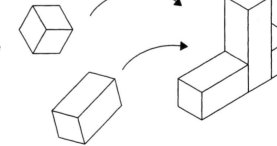

2. Use another identical set of these five rods to make another building and then compare it with your original. Can you make

   - a building with the same volume but more surface area?

   - a building with the same volume but less surface area?

   - a building with different volume?

   What is the greatest and the least surface area of buildings that you can construct with these rods?

3. Write about your strategies for finding the surface area and volume.

Fraction Rod Race

This is a game for two to four players. You will need a pair of regular dice and a set of 74 Cuisenaire Rods.

1. Each player should make a "race track" as follows. Draw a line segment the length of two dark-green rods. Label the points 0, 1 and 2 as shown.

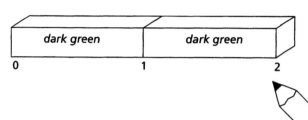

2. Use other rods to find and label the points $\frac{1}{2}$, $1\frac{1}{2}$, $\frac{1}{3}$, $\frac{2}{3}$, $1\frac{1}{3}$, $1\frac{2}{3}$, $\frac{1}{6}$, $\frac{5}{6}$, $1\frac{1}{6}$, $1\frac{5}{6}$.

3. Play the game. Take turns. On your turn, roll the dice and use the numbers shown to form a fraction. Use the smaller number as the numerator and the greater as the denominator. (If these numbers are the same, your fraction will be equivalent to 1.) If possible, take a rod that represents the fraction you roll and place it on your race track, starting at 0. If no such rod exists, your turn is over. On each turn, place the new rod next to the previous one to form a train. The object is to reach 2 exactly. If you roll a fraction that makes a train extending beyond 2, you must subtract a rod of that length from your train (exchanging rods if necessary).

4. Play the game at least two more times. Then write answers to these questions. From which positions do you have the best chance of winning on your next turn? From which positions is it impossible to win?

Tossing Four Counters

Work with a partner, taking turns tossing the counters, and record the results. You will need a graph like the one shown, four Two-Color Counters, and a container.

1. Put the counters in the container and shake it. Then spill the counters onto a flat surface. Count how many counters turn up red and record that number by putting an X on your graph. Do this once more and record your new results.

2. Predict which number will have the most X's if you spill the counters many times.

3. Now spill and record 30 more times.

4. When you are done, write your answers to these questions.

   • Did your prediction match your results?

   • Which number on the graph has the most X's?

   • Why do you think that happened?

   • How did your results compare to those of your classmates?

5. Do the Two-Color Counter activity *Arranging Four Counters*. Afterwards, compare your results from the two activities.

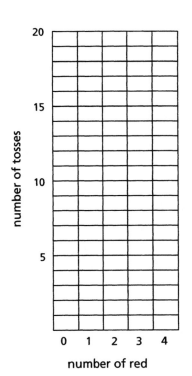

Arranging Four Counters

You will need four Two-Color Counters. (Do this activity after you have done *Tossing Four Counters*.)

1. In how many ways can you arrange the Two-Color Counters in boxes like these? Put only one counter in each box. *Note that the two arrangements to the right count as different.*

   • Record and organize your arrangements in a list.

   • Write about why you are sure that you have found all possible arrangements.

2. Complete a graph like the one to the right to show how many arrangements have no reds, 1 red, 2 reds, and so on.

3. Compare your graph to the one you made in *Tossing Four Counters*. How are the graphs alike? How are they different? If you made a graph showing the results of 1,000 spills of the counters, what do you think it would look like?

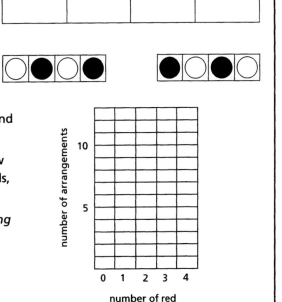

Fraction Flips

You will need Two-Color Counters.

1. Here are three riddles. One has no solution. Which one is it?

Riddle 1 I had some counters. One third of them were yellow. I flipped two of them over. Then one half of them were yellow. What counters did I start with?	**Riddle 2** I had some counters. One third of them were yellow. I flipped three of them over. Then one half of them were yellow. What counters did I start with?	**Riddle 3** I had some counters. One third of them were yellow. I flipped three of them over. Then two thirds of them were red. What counters did I start with?

2. Some riddles have many solutions. Consider this: I had some counters. One third of them were yellow. I flipped some of them over. Then one half of them were yellow. What can you say about the number of counters started with?

3. Using counters, make up your own fraction riddle for others to solve. Have someone try your riddle. Write it on an index card for a class set of riddles.

Collecting Counters

This is a game for two to four players. You will need two transparent spinners and spinner backings and about 80 Two-Color Counters.

1. Each of you begins with 4 yellow counters.

2. On your turn, spin both spinners and follow these rules:
   - A red and a yellow cancel each other out.
   - You can add or remove an equal number of reds and yellows at any time. For example, suppose you have 2 reds and spin "Give away 2 yellows" and "Take 1 yellow." To do this, you can take 2 yellow-red pairs from the bank, return 2 yellows, take 1 yellow, and then return 1 yellow and 1 red.
   - At the end of each turn, you must have the fewest number of counters possible—all reds or all yellows.

3. Score points according to which counters you have at the end of your turn:
   1 point for more than 10 yellow counters;
   2 points for 0 counters;
   3 points for 5 counters (of either color).

4. The first player to score 10 points wins.

5. Play the game at least three times. Write about the game. What is the fewest number of rounds a player needs to win? Does it make a difference in which order you use the spinners? If so, give an example. Do you think that the way points are assigned in the game is fair? Did it affect how you played?

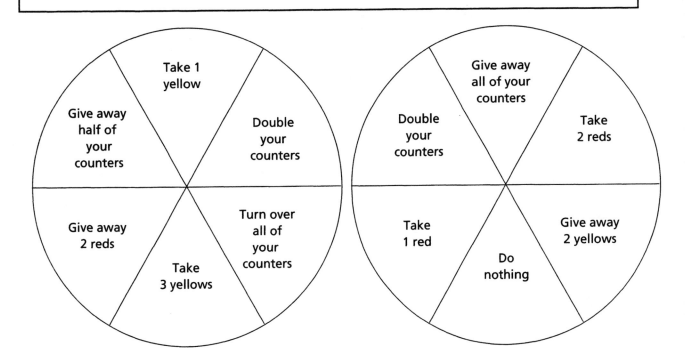

Chip Patterns on a Hundred Board

You will need transparent colored chips, a hundred board, and a calculator.

1. Put a blue chip somewhere on the hundred board, and surround it by 10 red chips as shown.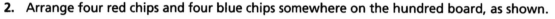

    Add the numbers covered by red chips. Repeat three more times.

    *   Write about what you notice and why you think it happens.

    *   Can you use an entirely different arrangement of ten red chips and one blue chip and get the same result?

2. Arrange four red chips and four blue chips somewhere on the hundred board, as shown.

    Add the numbers of each color. Repeat three more times.

    *   Write about what you notice and why you think it happens.

    *   Can you use a different arrangement of four red chips and four blue chips and get the same result?

3. Take nine red chips and make a square on the hundred board as shown. Position them so that the sum of the numbers covered will be each of the following:

162	360	378	596

    *   For each sum, write a description of your strategy or tell why you think that no such position exists.

Chips in a Row

This is a game for two to five players. You will need transparent color chips, a hundred board, four standard dice, and a calculator. Each player uses a different color of chip. You may want to use a timer.

1. Warm-up

   If you roll four dice, you will get four digits. These digits can be used together with any arithmetic operations to make a number. For example, suppose you rolled 2, 4, 5, and 6. Here are some numbers that you could make:

   $$0 = \frac{(4+6)}{2} - 5 \qquad 16 = 25\% \text{ of } 64 \qquad 19 = 2.5 \times 6 + 4$$

   $$46 = 2 \times (4 \times 5) + 6 \qquad 80 = 24 + 56 \qquad 100 = (4 + 6) \times (2 \times 5)$$

   Find ways to make at least four more numbers.

2. Now, play the game.

   - Take turns rolling all four dice.

   - On your turn, choose a number you can make with the four digits you have rolled and write it down. Cover this number with a chip.

   - Do not put a chip on a number that has already been covered.

   - The first player to get three of his or her chips in a row—horizontally, vertically, or diagonally—wins the game.

As a group, decide whether you want to use a timer to limit a player's time and, if so, how much time to give.

Hundred Boards

Cover by Clues

This is a game for two to five players. You will need transparent color chips, a hundred board, and an envelope with clue cards. Each person uses a different color chip.

1. Warm-up

   Take a clue card out of the envelope. Each card describes some numbers on the hundred board. Each description contains a blank to be filled in with a number or phrase from a list. As a group, decide which number or phrase describes the most numbers and which choice describes the fewest numbers. Repeat this process for a different card. Then, put the cards back in the envelope.

2. Now, play the game.

   • Take turns drawing a card from the envelope. The person with the clue card that describes the fewest numbers covers those numbers with a chip. Put that clue aside. Return the other cards to the envelope and pass it clockwise to the next player.

   • On your turn, draw a clue card, decide how to fill in the blank, and use a chip to cover all of the still uncovered numbers that the card describes. Put that clue card aside.

   • Play until all the clue cards have been used. The player with the most chips on the board is the winner.

3. List the numbers that remain uncovered. Write as few clues as possible that will cause these numbers all to be covered.

4. Write about the visual patterns created by the clues.

Multiples of ___ {5, 6, 7}	Prime numbers ___ {less than 50, greater than 40}
Multiples of ___ {8, 9, 11}	Units digits is___ {0, 2, 4, 6, 8}
Even numbers ___ {less than 30, greater than 70}	Tens digits is___ {0, 3, 5, 7, 9}
Odd numbers ___ {less than 30, greater than 64}	There is a___ {0, 3, 6, 9} in at least one digit
Sum of the two digits is ___ {7, 8, 9, 10, 11}	One digit is___ {the same as, double, triple} the other
Difference of the two digits is___ {1, 2, 3}	Both digits are ___ {less than 4, greater than 6}

Labels on Loops

You will need a set of labels, two loops of string, and 26 AlphaShapes of one color. Work with a partner.

1. Warm-up

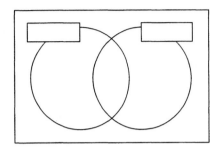

   - Pick one label and find all of the shapes that have that property. Repeat this with another label.
   - Make two overlapping loops of string, as shown. Pick two labels and put one on each loop. Place all of the AlphaShapes where they belong, inside or outside a loop or in the overlapping area.
   - Pick two more labels and do the same thing.

2. Now, play the game.

   - One person picks labels and places them facedown on the loops.
   - The other person guesses what is on the labels by placing a piece somewhere inside and waiting to hear whether the placement is correct. If the person who chose the labels says "yes," leave the piece in place. Otherwise, either move the piece or remove it and try putting down another piece.
   - Record your guesses each time. The object is to guess the labels with as as few tries as possible.
   - Play at least four times; switch roles each time.

All sides are congruent.	All angles are congruent.	There is a right angle.

There are three angles.	There are four sides.	The shape has line symmetry.

The shape has rotational symmetry.	Exactly two sides are parallel.	Opposite sides are parallel.

No sides are parallel.	No sides are congruent.	Opposite sides are congruent.

At least two sides are congruent.

AlphaShapes

Properties of Shapes

You will need 26 AlphaShapes of one color.

1. Below are some properties of a shape. Find all the AlphaShapes that satisfy *all* of these statements.

   (a) There are four sides.

   (b) There are at least two congruent sides.

   (c) There are four angles.

   (d) The shape has at least one line of symmetry.

   (e) The shape has a right angle.

   (f) There are two pairs of congruent angles.

   (g) All angles are congruent.

   (h) There is at least one pair of parallel sides.

   (i) There are two pairs of parallel sides.

   (j) Both diagonals cut the shape into two congruent parts.

2. Cover up one of the properties above so that the same set of AlphaShapes you found for part 1 still satisfies the remaining properties. Then cover every other property that can be covered without making the set change. Make a list of the properties that remain.

   Uncover all of the properties and repeat this, but try to do it in a different way so that a different list of properties remains. Compare your lists with those that others have made.

3. Take AlphaShapes K, O, Q, U, W. List all the properties that are shared by *all* these shapes. Make sure that no other AlphaShapes satisfy all of the properties you have listed. Then repeat part 2, using this list.

Ways to Find Area

You will need 26 AlphaShapes of one color, a transparent centimeter grid, and an angle ruler.

1. Compare the area of each piece to pieces W and Q. Sort the shapes into five piles, as shown.

area less than W	area same as W	area greater than W but less than Q	area same as Q	area greater than Q

2. Write about your results and explain how you compared areas.

3. Find the area in square centimeters of at least two shapes from each of the five piles. If you can't find the area exactly, estimate. Write about how you found the areas.

4. Find some AlphaShapes with the same area. Measure their perimeters. What do you notice?

Tracing Angles

You will need all of the polygons found among 26 AlphaShapes of one color.

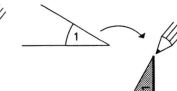

1. Take a triangle. The angles are numbered 1, 2, and 3.
   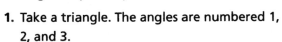
   - Trace around angle 1 and label the angle.
   - Place the shape so that the vertex of angle 2 is on the vertex of the angle you traced and a side matches.
   - Trace around angle 2 and label it.
   - Repeat with angle 3.

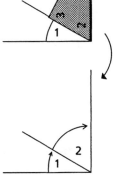

2. Try this for at least two more triangles. Describe in writing what you notice.

3. Try the same thing with some four-sided shapes. Now the angles are numbered 1, 2, 3, and 4. Describe in writing what you notice.

4. Predict what will happen for all the rest of the AlphaShapes and explain your thinking. Then check your predictions.

Tessellations

You will need a set of 52 AlphaShape pieces, some construction paper, and scissors.

Some of the AlphaShape pieces can be used to make tessellations—tiling patterns that completely cover a surface with no overlaps and no gaps. For example, S tessellates; F doesn't.

Tessellates   *Doesn't tessellate*

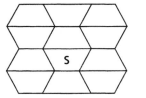

1. Separate all the AlphaShape pieces into two piles: shapes that can tessellate; shapes that can't tessellate. If you aren't sure where a shape belongs, cut out at least 10 construction paper copies and try to make a tessellation.

2. In what ways are the piles different? Explain in writing what makes only some shapes tessellate.

Circles

You will need at least five circles of different sizes (such as jar lids, coins, Two-Color Counters), 1 of AlphaShape X, 2 of AlphaShape Z's, string, an angle ruler, a compass, and the transparent centimeter grid.

1. Arrange all your circles, including the one made from the two Z's, in size order.
   - Estimate in centimeters, then measure the radius, diameter, and circumference of each circle. Record your findings.
   - Look for patterns and make a list of what you notice.

2. Draw or find a circle whose diameter is half as big as circle X. Use this circle along with circle X and the circle made from the two Z's to find the following information. Record your answers.
   - How do the measures of the radii compare?
   - How do the areas compare?
   - How does the measure of each circle's radius compare to its area?

3. Choose any of your circles. Make two new circles—one whose radius is twice as big and one whose radius is half as big as the circle you choose.
   - Predict the area of each circle.
   - Measure the area of each circle.
   - Explain, in writing, why your findings make sense.

Ways to Measure Angles

You will need a hinged mirror and 26 different AlphaShapes.

1. Pick a triangle, a quadrilateral, a pentagon, and a hexagon from the set of AlphaShapes. If you added up the measures of all of the angles of each shape, which would have the greatest sum? the least? Why do you think so?

2. Use one of the following methods to check your estimates. Record your findings.

   • Put AlphaShape V—a triangle—between the mirror faces so that two of its sides lie flush with the faces. How many copies of the shape can you see? A full circle contains 360°. Divide 360 by the V's that you see, including the original piece. This is the measure of the angle between the mirrors. Repeat for the other angles of V.

   • Put together some angles that you know. For example, the measure of any angle of shape W, a square, is 90°. The smaller angles (1 and 3) of triangle V measure 45°. (Placing angle 1 or angle 3 of V on top of any angle of the square shows this 1:2 relationship.) Now put a square and the smaller angle of triangle V on top of angle 2 of shape C. They just cover the angle. This means that this angle measures 90° + 45° = 135°.

3. Now pick a different set of four AlphaShapes. Include a triangle, a quadrilateral, a pentagon, and a hexagon.

   • Estimate and/or measure the angles and find the sum for each shape.

   • Compare your results to those of the first set.

   • Write about what you notice.

Lines of Symmetry

You will need 26 different AlphaShapes and the hinged mirror. Set the mirror at 180° for this activity.

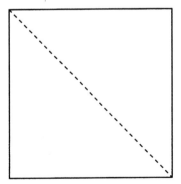

1. Place the mirror on the dotted line in this square. This dotted line is called a *line of symmetry*. A line of symmetry is a line along which you can place a mirror so that half of the original shape and its reflected image make the same image as the original shape. A square has four lines of symmetry. Find the other three.

2. Collect all the four-sided AlphaShapes. Determine how many lines of symmetry each one has. Agree or disagree with the following statements. Explain your reasoning.

   • A four-sided polygon has no more than four lines of symmetry.

   • A four-sided polygon can have any number of lines of symmetry up to four.

   • A four-sided polygon with four lines of symmetry has all of its sides and angles congruent.

3. How many lines of symmetry can a triangle have? a pentagon? a hexagon? Investigate and record your results. What patterns do you notice?

Finding Shapes

You will need a hinged mirror and 26 different AlphaShapes.

1. Put your hinged mirror on this drawing of shape J and move it around so that you see new shapes.

   • Find a way to make a square and a second rectangle that is not a square.

   • Use the mirror to make five more shapes and sketch each one.

2. Sort the following shapes into those that can be made by placing the hinged mirror on your figure and those that can't. Find a way to record the placement of the mirror so that you can discuss your results with someone else.

   rhombus (not a square)

   parallelogram (not a rectangle or a rhombus)

   trapezoid

   quadrilateral with no sides congruent

   kite (two pairs of congruent adjacent sides)

   equilateral triangle

   isosceles triangle (not equilateral)

   scalene triangle

   right triangle

   pentagon with all of its angles congruent

   pentagon with two right angles

   hexagon

   decagon

3. Why are some shapes possible and others not? Give examples.

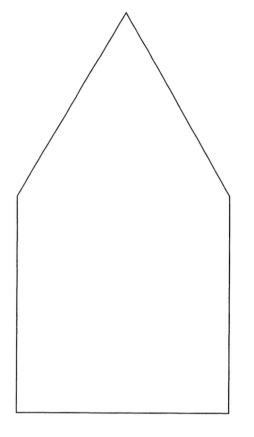

Geoboard Dot Paper—12 Squares

Teaching with Manipulatives: Middle School Investigations

ETA/Cuisenaire®

One–Centimeter Grid Paper

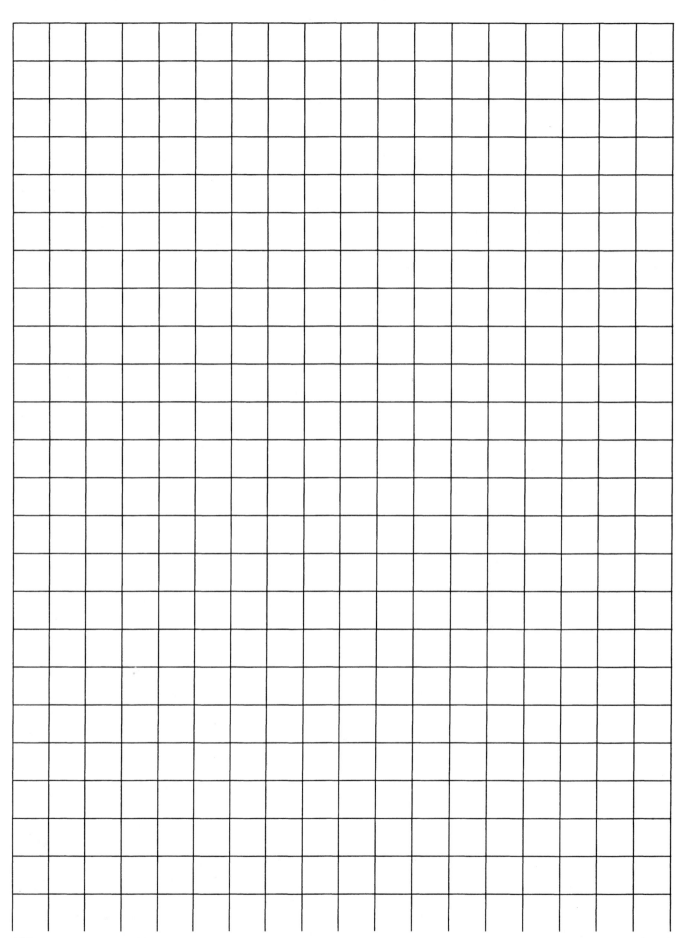